SpringerBriefs in Applied Sciences and Technology

Nonlinear Circuits

Series editors

Luigi Fortuna, Catania, Italy
Guanrong Chen, Heidelberg, Germany

SpringerBriefs in Nonlinear Circuits promotes and expedites the dissemination of substantive new research results, state-of-the-art subject reviews and tutorial overviews in nonlinear circuits theory, design, and implementation with particular emphasis on innovative applications and devices. The subject focus is on nonlinear technology and nonlinear electronics engineering. These concise summaries of 50–125 pages will include cutting-edge research, analytical methods, advanced modelling techniques and practical applications. Coverage will extend to all theoretical and applied aspects of the field, including traditional nonlinear electronic circuit dynamics from modelling and design to their implementation. Topics include but are not limited to:

- nonlinear electronic circuits dynamics;
- Oscillators;
- cellular nonlinear networks;
- arrays of nonlinear circuits;
- chaotic circuits;
- system bifurcation;
- chaos control;
- active use of chaos;
- nonlinear electronic devices;
- memristors;
- circuit for nonlinear signal processing;
- wave generation and shaping;
- nonlinear actuators;
- nonlinear sensors;
- power electronic circuits;
- nonlinear circuits in motion control;
- nonlinear active vibrations;
- educational experiences in nonlinear circuits;
- nonlinear materials for nonlinear circuits; and
- nonlinear electronic instrumentation.

Contributions to the series can be made by submitting a proposal to the responsible Springer contact, Oliver Jackson (oliver.jackson@springer.com) or one of the Academic Series Editors, Professor Luigi Fortuna (luigi.fortuna@dieei.unict.it) and Professor Guanrong Chen (eegchen@cityu.edu.hk).

Members of the Editorial Board:

More information about this series at http://www.springer.com/series/15574

Karabi Biswas · Gary Bohannan
Riccardo Caponetto
António Mendes Lopes
José António Tenreiro Machado

Fractional-Order Devices

 Springer

Karabi Biswas
Department of Electrical Engineering
Indian Institute of Technology Kharagpur
Kharagpur, West Bengal
India

António Mendes Lopes
UISPA–LAETA/INEGI
Faculty of Engineering, University of Porto
Porto
Portugal

Gary Bohannan
Department of Physics and Materials
  Science
University of Memphis
Memphis, TN
USA

José António Tenreiro Machado
Department of Electrical Engineering
Institute of Engineering
  of Polytechnic of Porto
Porto
Portugal

Riccardo Caponetto
Department of Electrical, Electronics and
  Computer Engineering
University of Catania
Catania
Italy

ISSN 2191-530X   ISSN 2191-5318   (electronic)
SpringerBriefs in Applied Sciences and Technology
ISSN 2520-1433   ISSN 2520-1441   (electronic)
SpringerBriefs in Nonlinear Circuits
ISBN 978-3-319-54459-5   ISBN 978-3-319-54460-1   (eBook)
DOI 10.1007/978-3-319-54460-1

Library of Congress Control Number: 2017933432

Printed on acid-free paper

This Springer imprint is published by Springer Nature
The registered company is Springer International Publishing AG
The registered company address is: Gewerbestrasse 11, 6330 Cham, Switzerland

Acknowledgements

Karabi Biswas would like to acknowledge Prof. Siddhtartha Sen, Professor, Electrical Engineering Department, IIT Kharagpur, Dr. Madhab Chandra Tripathy, Dr. Munmun Khanra, Mr. Avishek Adhikary, Ms. Debasmita Mondal, (graduate and undergraduate students who worked with me) and Ms. Dina Anna John and Mr. Pratyush Prakash who reviewed the draft.

Gary Bohannan would like to acknowledge Dr. Melissa McIntyre, Dr. Cal Coopmans, and Dr. Stephanie Hurst for their efforts in developing and demonstrating the early fractance devices, Mr. Andrew Skytland for developing the second-generation fractance devices, and Ms. Brenda Knauber for assistance in development of the fractional harmonic oscillator and for her review of early drafts.

Riccardo Caponetto would like to acknowledge Prof. Luigi Fortuna for addressing the study of fractional-order system and for the continuous support and encouragement.

Contents

1 Introduction to Fractional-Order Elements and Devices 1
   1.1 Introduction 1
   1.2 Historical Background 1
   1.3 Motivation 3
   1.4 What is a "Fractance" or "Fractional-Order Device"? 4
   1.5 The Case for Power–law ("Fractional") Dynamics 5
   1.6 A Brief Introduction to the Fractional Calculus 9
   1.7 The Push for a New Electronic Device 17
   References 19

2 Devices 21
   2.1 Introduction 21
   2.2 Discrete Element Approximations of Fractional-Order
        Elements 22
   2.3 Early Fractional-Order Devices (FOD) 23
        2.3.1 Platinum Nanowire in Polymer 23
        2.3.2 Lithium-Ion Type 25
        2.3.3 Modified Lithium-Ion Type 28
        2.3.4 Other Lithium-Ion Attempts 29
   2.4 Nanostructured Materials as Fractional-Order Elements 30
        2.4.1 IPMC Structure and Working Principles 30
        2.4.2 Linearity Study 33
        2.4.3 Carbon Black-Based FOE 39
        2.4.4 FOE Under Test and Experimental Setup 39
   2.5 Other Solid-State Devices 42
   2.6 Solution-Based Systems 44
        2.6.1 Fabrication Details of the Solution-Based FOE 45
        2.6.2 The Parameter Dependence of the FOE 48
        2.6.3 CNT–Polymer Composite-Based Wideband FOE 50
        2.6.4 Fractance-Based Sensor 50

2.7   Lesson Learned   51
References   51

3   Demonstrations and Applications of Fractional-Order Devices   55
3.1   Introduction   55
3.2   Circuit and System Design Using Fractional-Order Elements.   55
3.3   Control System Demonstrations   56
       3.3.1   Temperature Control Demonstration   56
       3.3.2   Robotics Control Demonstration.....................   57
3.4   Cascaded Circuit Demonstration   60
3.5   Circuit Demonstrations Using Solution-Based FOE   62
       3.5.1   Filter Circuit Demonstration   65
       3.5.2   PLL Circuit Demonstration.........................   67
       3.5.3   Resonator Circuit Demonstration   69
3.6   Conclusion   71
References   72

4   Fractional-Order Models of Vegetable Tissues   73
4.1   Introduction   73
4.2   Empirical Fractional-Order Models   75
4.3   On the Fractional-Order Models of Vegetable Tissues   79
4.4   Modeling Different Size Stems of a Plant   79
4.5   Modeling Fruits and Vegetables...........................   80
4.6   Clustering and Visualizing   85
4.7   Conclusions   89
References   89

5   Future Directions   93
5.1   Introduction   93
5.2   Challenges and Opportunities.............................   93
5.3   Achieving Specific Fractional-Order and Longer
       Working Lifetimes   94
5.4   Advanced Research Opportunities   94
5.5   Dynamic Fractance and Memfractance......................   94
5.6   Generalizing Ohm's Law   96
5.7   Teaching Fractional Calculus and Its Applications.............   99
5.8   Conclusion   100
References   101

Abbreviations and Symbols

Abbreviations

C	designation for Caputo fractional-order operator or capacitance. These are recognized in context
CC	Cole–Cole model for constant phase element behavior
CD	Cole–Davidson model for more complex constant phase element behavior
CPE	constant phase element, mathematical model for describing specific impedance properties
CPZ	constant phase zone, the frequency band over which the item shows constant phase
D	Debye model for dielectric behavior
FC	fractional calculus, the generalized calculus of arbitrary order
FO	fractional order
FOD	fractional-order device, a device created from material exhibiting fractional-order dynamics
FOE	fractional-order element, a material exhibiting fractional-order dynamics
GL	designator for the Grünwald–Letnikov fractional-order operator
HN	Havriliak–Negami model for complex constant phase element behavior
IES	electrical impedance spectroscopy
PL	power law
RL	designation for the Riemmann–Liousville fractional-order operator

Symbols

α, β	the fractional-order exponents
ε	dielectric permittivity function
f	frequency of applied stimulus signal
i	current flowing through the device
j	the imaginary radix $\sqrt{-1}$

ω	angular frequency = $2\pi f$
v	voltage measured across a device
$_aI_t^\alpha$	the fractional-order integral over the interval [a,t], often also designated by C or RL to denote the specific definition being used
Z	frequency-dependent electrical impedance of a device

Chapter 1
Introduction to Fractional-Order Elements and Devices

1.1 Introduction

In this chapter, we begin with an outline some of the history behind the development of the early fractional-order (FO) devices (FOD) along with the evolution of the motivations for their creation. This inevitably brings up a discussion of the calculus of arbitrary order, the fractional calculus (FC), that best models the behavior of these devices. We will also introduce some of the methods of impedance spectroscopy to describe how the behavior is measured. We will attempt to bring several seemingly disparate aspects of mathematics, material science, and electronic circuit design into a coherent focus on developing a new class of devices.

In Chap. 2, we will introduce some of the early implementations of these new devices. No attempt will be made to survey all of the implementations of FO elements and devices; there are just too many in development at the time of this writing. Rather, we intend to introduce some ideas of how to get start making more devices and exploring new applications.

In Chap. 3, we will introduce methods for designing circuits incorporating FOD and review some of the demonstrations of applications of these devices. In Chap. 4, we will look deeper into nature's versions of FO materials. Finally, we will look the future of research in fractional-order dynamics; how can we use FOD to explore some open questions in complex system dynamics.

1.2 Historical Background

The "calculus of arbitrary order" began with a letter between L'Hopital and Leibnitz over the meaning of $d^\alpha x / dx^\alpha$, with α being $1/2$. It was considered a paradox that would someday lead to useful applications. Interest in such matters came and went sporadically over the years with developments by Grünwald and Letnikov, Reimann, Liouville, and a few others. Oldham and Spanier published the first actual text on the subject in [21]. Their contributions will be discussed in more detail in subsequent sections. The point here is that this was mainly from a purely mathematical viewpoint.

© The Author(s) 2017
R. Caponetto et al., *Fractional-Order Devices*,
SpringerBriefs in Nonlinear Circuits, DOI 10.1007/978-3-319-54460-1_1

The applications they discussed had been solved by conventional methods, so did not lead to sustained interest.

On a parallel track, there was interest in noninteger order control beginning in the middle of the twentieth century with H. Bode's seminal work inventing feedback circuits and control theory [2]. Without actually mentioning fractional-order control (FOC), Bode implemented a 19 element circuit to achieve a desired 60° phase control circuit for use in signal processing. This was important enough that Bode abandoned two additional chapters he had intended to include in his text of 1945. Ironically, just a few years earlier Cole and Cole discovered the power–law (PL) behavior in numerous materials and gave this characteristic the label "constant phase element" (CPE) [11]. Examples of early work on FOC include efforts by Manabe and Carlson. Manabe used the ideas of noninteger order control to stabilize autonomous flight dynamics [19]. Carlson had developed an approximation of a $1/2$ order integrator circuit using a ladder network of resistors and capacitors and simulated its operation in 1961 as part of his graduate program [7]. Notable in that work is the reference to $1/2$ order control being optimal for diffusion limited systems such as nuclear reactors. Further detailed discussion on this will be the subject of Chap. 3.

The situation began to change in earnest in the 1990s. By the end of the millennium, the connections between the mathematics and the physical properties of materials and the possible applications to control theory and signal processing began growing exponentially. By 1995, the idea of FOC had been discussed by Machado in technical meetings [30], and later published in as a journal article [31]. I. Podlubny included the concepts of FOC design in his text on fractional differential equations [25]. The epiphany was that integer order models of complex system performance were proving to be inadequate. Not all systems to be controlled obeyed the assumptions of linear, lumped, time invariant models. While numerous texts had been published over the years describing compensation techniques to get around the limitations of simple controls made up from proportional, integral, and derivative operations, design and tuning to incorporate these techniques was cumbersome and costly. These techniques involved designing complex analog circuits or digital code to implement "lead/lag" phase compensation, often with limited performance improvement.

At a workshop on FOC, held in Las Vegas, at the end of 2001, a proposal was put forward to simplify the design of fractional-order controllers by creating a single circuit element that had a nearly constant phase angle of approximately 45° over a broad band of frequencies [4]. A single crystal of Lithium Hydrazinium Sulfate ($LiN_2H_5SO_4$) had been studied and the description included the comment that the real and the imaginary parts of the dielectric function were equal and varying as $f^{-1/2}$ [29]. As will be discussed in detail in Sect. 1.5, this implied that the $LiN_2H_5SO_4$ exhibited the $1/2$ order impedance associated with diffusion of charge carriers through the crystal. A device with this characteristic could potentially be used in an analog feedback circuit to create fractional-order mathematical function.

Such nearly pure single exponent PL dielectric function had not been recognized as having value. Whether such behavior had been witnessed before is unknown, but no discussion of it had been included in references on impedance spectroscopy. It was generally assumed that a nearly pure single exponent PL impedance was not possible. The "constant phase element" property had always been masked by other effects. In any case, other $LiN_2H_5SO_4$ characteristics made such crystals unsuitable for direct application in electronic circuits. With knowledge that nearly pure PL impedance was possible, Bohannan suggested a research project to deliberately fabricate materials that exhibited PL impedance, suggesting the term "fractance," short for fractional impedance. The goal being to demonstrate a device made from this new material in a control system. The project proceeded with no external funding, but proved successful in less than 3 years. The results will be discussed in Chap. 2. Results of demonstrations using these devices will be discussed in Chap. 3.

Meanwhile, efforts continued on developing analog and digital approximations of fractional-order impedance. Until such a time as a true fractional-order device was available, the only way to implement a fractional-order operator in a circuit was to synthesize it from conventional elements, such as the ladder circuit proposed by Carlson. Other synthesizing techniques followed [6, 8, 9, 26]. Numerical methods were also being developed and demonstrated in control system applications [22, 23, 32]. Some of these techniques will be reviewed in subsequent chapters for comparison with actual fractional-order devices.

For those not familiar with the concepts of the fractional calculus, the next sections of this chapter are devoted to outlining the concepts and motivations behind its study. We will attempt to tie together a number of different viewpoints, including how the description of the dynamics of fractance devices using the fractional calculus allows us to turn the description around and use the devices to build mathematical operators. For more on the history of the development of the fractional calculus, there is an abundance of literature, see, for example [33, 38].

We will then turn to a deeper discussion of impedance spectroscopy and interpreting the data obtained. This will allow us to build working models using the fractional calculus and then predict the performance of fractional-order circuits.

1.3 Motivation

At the Las Vegas workshop on fractional-order control theory in 2001, as well as elsewhere, the conjecture was put forward that FOC could perform better than the best integer order control. While this conjecture is still prevalent and there are several demonstrations suggesting some truth in the statement, it has not been rigorously proven.

While control system applications appear to have the most immediate practical payoff, there are other, deeper, motivations for pursuing development of FOD.

Numerous proofs have been presented that the simple exponential response predicted by purely imaginary impedance functions such as that of an "ideal" capacitor, $^1/_{cs}$, violate causality. See, e.g., Jonscher's text [18]. (The proof is beyond the scope of this monograph, but invokes the Kramers–Kronig relation used extensively in the study of optical spectroscopy.) A different proof was presented by Sakurai in a graduate level quantum mechanics text [28].

What then was the "correct" description of capacitive impedance? After testing thousands of capacitors of every kind they could obtain, Westerlund and Ekstam came to the conclusion that it should be written as a power–law relation [43].

But why power–law? There had been extensive evidence of PL as the most common description of complex dynamics of all kinds. Suggestions for why PL come from a generalization of the law of large numbers. Under the assumption that some observable made up of the sum of a very large number of identical independent actions, such as currents in electrical devices being made up of the combined effects of a large number of charge carriers in motion, the sum distribution will exhibit the Gaussian, or bell curve, also described as the Normal distribution. What happens when there are interactions that violate the assumption of independence? The second moment of the distribution fails to converge. That is, there is no standard deviation; it tends towards infinity. In real-time sampling, the mean might converge rapidly, while the estimate of the standard deviation fluctuates radically. In this case, the Generalized Law of Large Numbers takes over. The result is PL behavior, typified by what is known as the α-stable distribution [16]. This is a very general statement and suggests that power–law is "the law" for complex systems. Numerous reviews the ubiquity of PL distributions have been published, see for example [24, 35, 40]. The connection between PL statistical distributions and fractional dynamics was included in the second volume of Feller's text on statistics [13]. The result of all this is the expectation of fractional-order dynamics in complex systems. The ubiquity of power–law relations was highlighted in Machado's article on chromosome codes, just as an example [34]. B. J. West summarized this in his treatise *Fractional Calculus View of Complexity: Tomorrow's Science* [41]. We will have more to say on this in subsequent chapters.

Much is left to be discovered. What is it about the interactions that brings out the specific PL exponent in any given case? Can we design test cases, such as electronic devices with predetermined exponents? There is interest in developing such devices, given that there is already at least one patent issued for determining the desired exponent to be used in a control system problem [10].

We now turn to a more in depth description of what is meant by a "fractional-order device" and how that relates to the fractional calculus.

1.4   What is a "Fractance" or "Fractional-Order Device"?

A fractance is an element whose impedance follows a power–law in frequency which can be written in several equivalent forms such as

$$Z(\omega) = \frac{F}{(i\omega)^{\alpha}} = \frac{K}{(i\omega/\omega_C)^{\alpha}}, \tag{1.1}$$

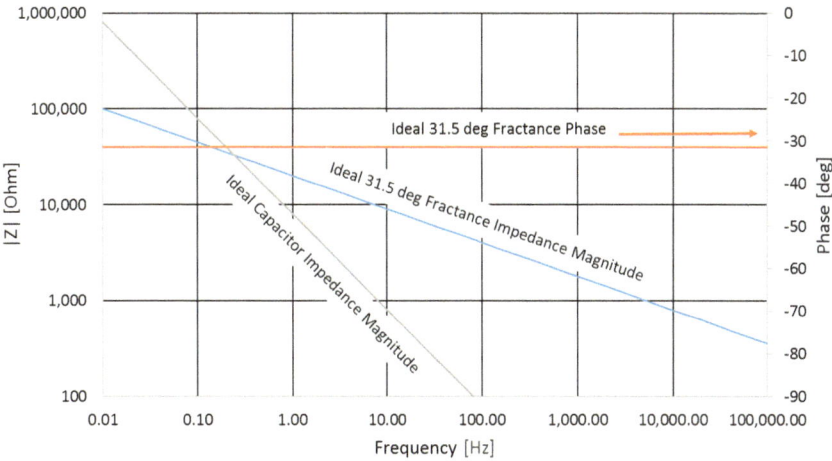

Fig. 1.1 Idealized impedance plot for a constant phase element of order $\alpha = 0.35$ giving a phase angle of $-31.5°$. The value of K is 4,000 Ω referenced at $f_C = 100$ Hz. The value of $\tau = 1/(2\pi f_C)$

with the exponent α being the "order" of the element and ω_C being the reference frequency, usually in the logarithmic middle of the frequency band of interest, for measuring the impedance magnitude K in ohm. Impedance magnitude plots are most often shown in log–log format as the values of $|Z|$ will show up as a straight line on such plots. The phase is usually shown on semi–log to allow visual connection between phase and magnitude at any given frequency. An ideal fractance response is illustrated in Fig. 1.1.

When one makes the generalization from a Fourier to Laplace description, one gets

$$Z(s) = \frac{F}{s^\alpha} = \frac{K}{(\tau s)^\alpha}, \quad \text{with } \tau = 1/\omega_C. \tag{1.2}$$

This description allows for resistance with $\alpha = 0$, ideal capacitance with $\alpha = 1$, and ideal inductance with $\alpha = -1$. Thus, any device described by (1.1), with a nearly constant phase $-\alpha\pi/2$, with arbitrary α, could be considered a fractance. In terms of the Laplace transform, $1/s$ can be interpreted as the first order integral, so $1/s^\alpha$ can be interpreted as the FO integral of order α. At this point, a bit of introduction to the calculus of arbitrary order, the "fractional calculus," is appropriate.

1.5 The Case for Power–law ("Fractional") Dynamics

At this point it is appropriate to introduce some of the terminology of impedance spectroscopy. This will be the briefest possible introduction. We will discuss more complex models in Chap. 4. See, for example, [1] for more details.

Engineers typically use impedance which is a property of a complete device. It represents the ratio of the voltage to the current through the device. Impedance properties are normally measured with respect to the response to a sinusoidal excitation. For a given frequency, $\omega = 2\pi f$,

$$v(t) = V\cos(\omega t + \theta_V), \tag{1.3}$$
$$i(t) = I\cos(\omega t + \theta_I), \tag{1.4}$$

where V and I, as well as θ_V and θ_I, can be frequency dependent. Expressing these in the frequency domain,

$$\mathbf{V}(j\omega) = V(\omega) \cdot e^{j\theta_V(\omega)}, \tag{1.5}$$
$$\mathbf{I}(j\omega) = I(\omega) \cdot e^{j\theta_I(\omega)}, \tag{1.6}$$

where $j = \sqrt{-1}$, the formal description of the complex impedance is then $\mathbf{Z}(j\omega)$ is defined as the ratio of phasors:

$$\mathbf{Z}(j\omega) = \frac{\mathbf{V}(j\omega)}{\mathbf{I}(j\omega)} = \frac{V(\omega)}{I(\omega)} \cdot e^{j\arg(\theta_V(\omega)-\theta_I(\omega))} = |\mathbf{Z}(j\omega)| \cdot e^{j\arg[\mathbf{Z}(j\omega)]}. \tag{1.7}$$

We will typically suppress the explicit dependence on frequency and just display the voltages and currents as V or v and I or i, with the capitalization referring to the amplitude of the variable under ac conditions. In most measurement scenarios, either the voltage or the current amplitude and phase are kept constant, but this is not strictly necessary. Another popular formal description for impedance is

$$\mathbf{Z} = \mathbf{R} + j\mathbf{X}, \tag{1.8}$$

where $\mathbf{R}$ is the real, or resistive, component of the complex impedance and $\mathbf{X}$ is the imaginary, or reactive, component. Again, we will follow common usage of just using the symbol Z for the frequency dependent impedance.

Materials scientists prefer to discuss the intrinsic properties of the dielectric material itself. The dielectric function, or permittivity, is obtained from impedance measurements by

$$\varepsilon = \varepsilon' + j\varepsilon'' = \frac{1}{j\omega C_C Z}, \tag{1.9}$$

where C_C is the capacitance of the empty test cell used in the measurements. In this form, ε' represents the energy storage component and ε'' the energy loss component. For historical reasons, the dielectric function is often written in its complex conjugate form as $\varepsilon^* = \varepsilon' - j\varepsilon''$. As used here, the "permittivity," or sometimes, the "dielectric constant," is formally the *relative* permittivity.

Impedance spectroscopy is the measurement of V versus I over a range of excitation frequencies. Numerous graphical plots are used, $|Z|$ versus f in log–log format

Fig. 1.2 Real (energy storage ε') and imaginary (energy dissipation ε'') components of the Debye model

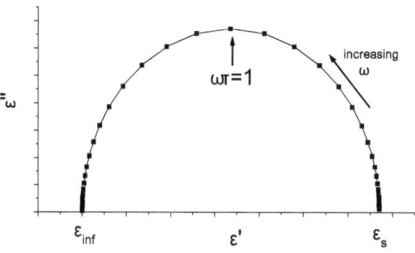

along with arg(Z) in log–linear format will be used regularly in this brief. Other popular formats include R versus X, ε'' versus ε' and many more. Each plotting format emphasizes a particular aspect of the material or device.

With this background, we turn to Debye's early model derived from measurements with dilute solutions of polar molecules. In terms of the complex conjugate of the permittivity $\varepsilon^* = \varepsilon' - i\varepsilon''$, simple polarization relaxation would appear as a semicircle on a plot of ε'' versus ε' [12]. Written in traditional form describing the polarization in a material due to an externally applied electric field:

$$\mathbf{P}(t) + \tau \frac{\partial}{\partial t}\mathbf{P}(t) = \varepsilon_s \mathbf{E}(t) + \varepsilon_\infty \tau \frac{\partial}{\partial t}\mathbf{E}(t), \tag{1.10}$$

$$\varepsilon^* - \varepsilon_\infty = \frac{\varepsilon_s - \varepsilon_\infty}{1 + (j\omega\tau)}. \tag{1.11}$$

An ideal equivalent circuit exhibiting the response of Fig. 1.2 would be as shown in Fig. 1.3.

Cole and Cole noticed that Debye's model for dielectric permittivity in dilute solutions did not hold for more concentrated solutions [11]. They found that for nondilute solutions, there was an effect due to interactions of some kind, leading to a PL correction to the description. The PL was deduced from curve fitting the permittivity plot, Fig. 1.4, and not from any first principle argument. They deduced the power–law exponent α from the geometry of the depressed center of what should

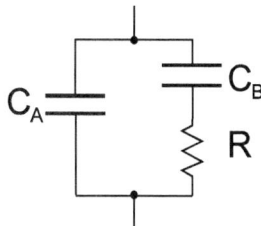

Fig. 1.3 Equivalent circuit of the Debye model. The model attempts to take into account the plate capacitance in the absence of the dielectric as well as the blocking effect of the electrode/electrolyte interface

Fig. 1.4 An idealized plot of the real and imaginary components of the dielectric "constant"

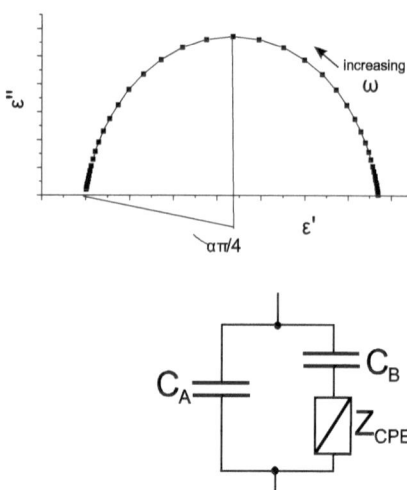

Fig. 1.5 Idealized equivalent circuit proposed by Cole and Cole [11] with Z_{CPE} depicting the constant phase element properties embedded in the dielectric response

have been a circle, but instead an arc, which could be best fit with a PL modification of Debye's model.

$$\varepsilon^* - \varepsilon_\infty = \frac{\varepsilon_s - \varepsilon_\infty}{1 + (j\omega\tau)^{1-\alpha}}. \tag{1.12}$$

They created the term CPE for the PL term. Note that the overall results of the measurements typically did not show constant phase, since the CPE was embedded with other processes. An idealized equivalent circuit that would exhibit this overall behavior is shown in Fig. 1.5. We can roughly identify C_A as the empty cell capacitance due to the electrode plates, equivalent to C_C above, and C_B as a blocking layer between the electrodes and the electrolyte exhibiting the CPE response. This PL response becames recognized as universal and is now commonly used in dielectric spectroscopy [1, 17].

Converting from permittivity (ε) to impedance by $Z = (j\omega C_A)^{-1}\varepsilon^{-1}$, the Cole–Cole model can be described by an idealized equivalent circuit shown in Fig. 1.5.

$$Z_{CPE} = Z_{Fractance} = \frac{K}{(\tau s)^\alpha}, \tag{1.13}$$

where we have made the generalization of $j\omega \rightarrow s$. The parameter K is the reference impedance, in ohm, at the reference frequency ω_C. The time scaling constant is derived from the reference frequency, $\tau = 1/\omega_C$. The order α is taken from the slope of the impedance magnitude on a log–log plot. It can also be derived from the average phase $\alpha = -\phi/90°$. We can associate ε_∞ with C_A and ε_s with $C_A + C_B$.

In parallel activity, researchers looking to characterize capacitors seemed surprised to find that "dead matter has memory" [42]. Apparently, they were unaware of the common universal PL nature of dielectrics. Westerlund and Ekstam developed the "fractional capacitor" impedance description in common use by engineers today [43].

$$Z_F = \frac{F}{s^\alpha} \qquad (1.14)$$

with $F = K/\tau^\alpha$ from (1.13) above. Note that F has fractional-order units but is accepted as an alternative to (1.13) above and both will be used throughout the rest of this brief.

Many paths have converged on the realization that PL impedance behavior is not only common, but truly universal. So, why not deliberately create a device with an exponent far from integer order? Fractional-order impedance had been observed as far back as 1899 by Warburg [39]. While the Warburg impedance theory admits only to $1/2$ order using the standard diffusion model, it did offer a suggestion for developing such a device. Further evidence of Warburg impedance and implicating Lithium as a possible charge carrier came from references such as [29]. Typically, however, the CPE behavior is mixed in with other dynamics as modeled by the equivalent circuit of Fig. 1.5.

The remaining question of whether we could depend on achieving power–law response was addressed by a statistical argument. With "adequate" randomness creating a very broad distribution of trapping and regeneration sites, the generalized law of large numbers dictate that the sum distribution, i.e., the observed averaged currents would exhibit a PL relation. The challenge was to create this very broad scale randomness and would that be enough? The experimental question was whether deterministic or random fractal structures might be required. After all, a single fractional-order impedance had not been identified as such up to the end of the 1990s. Another hint came from a study of transfer across random versus fractal interfaces [14] in which they detected no noticeable difference between deliberate deterministic fractal interfaces and heavily randomly roughened surfaces.

The connection between randomness and PL dynamics described by the fractional calculus had been discussed in articles such as [15]. Power–law CPE behavior has typically been described as exhibiting a broad distribution of exponential time constants due to the randomness of interactions. Why this distribution is so common has not been generally appreciated. Given the argument in the preceding paragraph, more sophisticated thinking might suggest that a PL description is actually more physical. The next step would be to work on making a device with randomized structure rather than the near "ideal" structures being employed in conventional devices such as low-loss capacitors. We will pick up this thread in Chap. 2.

1.6   A Brief Introduction to the Fractional Calculus

Differentiation and integration to arbitrary order, "the fractional calculus," had its origins with the very beginning of the development of the calculus itself. The meaning of a $1/2$ order derivative was considered paradoxical and remained a mathematical oddity for almost 300 years, with a few notable instances of interest springing up here and there. A detailed history up to 1974 was included in the text by Oldham and Spanier [21]. Since the mid-1990s, a considerable library of texts on the subject of fractional calculus and its applications has been published.

Since the fractional calculus is the generalization of integer order calculus, several different definitions have been proposed. Each of the definitions converge continuously to the integer order forms as the order approaches an integer value. It should be no surprise that there could be many possible forms with this characteristic. One of the objectives of studying fractance behavior is to help determine what form(s) accurately describe and predict natural processes. There is also the expectation that spatial forms of the FO operators may be quite different from those involving fractional-order in time. Each of the noninteger forms is "nonlocal" in the sense that time based operators, even derivatives, always include past history, not just the previous instant. Spatial operators account for nonnearest neighbor interactions. (The distinction between spatial and temporal operators is an active area of significant interest in the study of special and general relativity where space and time must be treated equivalently.)

Here we focus on the definitions that have been demonstrated to describe and predict physical time-dependent processes. The most used forms for temporal operators are the integral definitions by Riemann and Liouville (RL) and Caputo (C) and the limit sum form devised by Grünwald and Letnikov (GL).

All of these forms use the gamma function $\Gamma(\cdot)$ which the generalized factorial function.

$$\text{Definition}: \Gamma(x) = \int_0^\infty t^{x-1} \exp(-t)dt \quad x > 0 \tag{1.15a}$$

$$\text{Recursion Formula}: \Gamma(x+1) = x\Gamma(x) \tag{1.15b}$$

$$\Gamma(n+1) = n! \quad \text{if } n = 0, 1, 2, \ldots \text{ where } 0! = 1 \tag{1.15c}$$

$$\text{For negative } x: \Gamma(x) = \frac{\Gamma(x+1)}{x} \tag{1.15d}$$

The Riemann–Liouville (RL) form takes the integer order derivative of the FO integral.

$$_a^{RL}D_t^\alpha f(t) = \frac{1}{\Gamma(n-\alpha)} \frac{d^n}{dt^n} \int_a^t \frac{f(t')}{(t-t')^{\alpha-n+1}} dt', \quad n-1 < \alpha < n \tag{1.16}$$

The Caputo form takes the FO integral of the integer order derivative.

$$_a^C D_t^\alpha f(t) = \frac{1}{\Gamma(n-\alpha)} \int_a^t \frac{f^{(n)}(t')}{(t-t')^{\alpha-n+1}} dt', \quad n-1 < \alpha < n \tag{1.17}$$

The GL limit sum form allows for any value of α, including complex. Note that differential and integral calculus can be unified in a single definition. Interpretation of a complex order operator in terms of physical properties is an active area of research beyond the scope of this treatise.

$$
{}^{GL}_{a}D^{\alpha}_{t}f(t) = \lim_{N\to\infty}\left[\frac{t-a}{N}\right]^{-\alpha}\left\{\sum_{k=0}^{N-1}\frac{\Gamma(k-\alpha)}{\Gamma(-\alpha)\Gamma(k+1)}f\left(t-k\left[\frac{t-a}{N}\right]\right)\right\}
\tag{1.18a}
$$

or

$$
{}^{GL}_{a}D^{\alpha}_{t}f(t) = \lim_{N\to\infty}\Delta t^{-\alpha}\left\{\sum_{k=0}^{N-1}w_{k}\,f\left(t-k\Delta t\right)\right\}
\tag{1.18b}
$$

where the increment is $\Delta t = (t-a)/N$, which implies dividing up the interval into ever smaller subintervals. In practice, for numerical computation using the GL limit sum form, we can take advantage of the recurrence relations of (1.15) to build the weighting terms

$$
w_{k} = \frac{\Gamma(k-\alpha)}{\Gamma(k+1)\Gamma(-\alpha)} = \left(\frac{k-1-\alpha}{k}\right)\frac{\Gamma(k-1-\alpha)}{\Gamma(k)\Gamma(-\alpha)}
\tag{1.19a}
$$

or

$$
w_{k} = \left(\frac{k-1-\alpha}{k}\right)w_{k-1}\qquad\text{where } w_{0}=1.
\tag{1.19b}
$$

The Laplace transform of the Riemann–Liouville (RL) form:

$$
\mathcal{L}\{{}^{RL}_{a}D^{\alpha}_{t}f(t)\} = s^{\alpha}\mathcal{L}\{f(t)\} - \sum_{k=0}^{n-1}{}^{RL}_{0}D^{\alpha-k-1}_{t}f(a^{+}),
\tag{1.20}
$$

where a^{+} refers to the instant immediately after $t=a$.
   The Caputo (C) form:

$$
\mathcal{L}\{{}^{C}_{a}D^{\alpha}_{t}f(t)\} = s^{\alpha}\mathcal{L}\{f(t)\} - \sum_{k=0}^{n-1}s^{\alpha-k-1}f^{(k)}(a).
\tag{1.21}
$$

Laplace transform tables have included FO items for as long as they have been published. For example,

$$
\frac{\Gamma(k)}{s^{k}} \overset{\mathcal{L}}{\Longleftrightarrow} t^{k-1}\qquad k>0.
\tag{1.22}
$$

The difference between the Riemann–Liouville (RL) and Caputo (C) definitions is the order of operations when $n\neq 0$. In each case, the FO integral is of order between -1 and 0. The major difference is in the way initial conditions are handled, but as will be shown, this turns out to be irrelevant for physical systems. They end up giving the same results.

Note that the basic FO operators return results in fractional-order units. In the case of the FO time integral, the result includes a term involving $\sec^\alpha$. This is never a problem in real systems as there is always a scaling term with the same units to give a result in real integer order units. This term is the τ^α in the FO impedance description of (1.2).

Again, we emphasize that generalized derivatives, as well as integrals, are defined over a nonvanishing interval $[a, t]$. The interval may extend back to minus infinity. To obey causality, the interval cannot include future time. Experimentation gives us clues as to how to interpret the operators under initialized conditions. For example, the charge accumulated across an ideal capacitor is given by the complete integral of the current flowing through it:

$$q(t) = \int_{-\infty}^{t} i(t')dt' \tag{1.23}$$

which can be split up as

$$q(t) = \int_{-\infty}^{a} i(t')dt' + \int_{a}^{t} i(t')dt'. \tag{1.24}$$

Generalized derivatives have positive α, and generalized integrals have negative α, so we can write the anti-derivative as $_aI_t^\alpha =_a D_t^{-\alpha}$, where α is the order of the operator. Both derivatives and integrals are computed over a finite interval $[a, t]$. While we typically write equations of dynamics as differential equations, actual physical processes evolve as sequence of actions integrated over time. For this reason, to maintain causal relations, differential equations need to be recast as integral equations. The rule being to put the highest order derivative on the left-hand side and everything of lower order on the right-hand side. Assuming the highest order is α, carry out an α order integral on both sides to predict the time evolution of the process. This step of invoking causality is often ignored in integer order calculations, but is critical when taking the effect of prior history into account in FO systems. The initial condition $q_0 = \int_{-\infty}^{a} i(t)dt$ allows us to write

$$q(t) = \int_{a}^{t} i(t')dt' + q_0. \tag{1.25}$$

This seems trivially obvious at first, until one looks at the actual response of real systems that are best described by FO operators. A toy example illustrates the point. Consider a "leaky" capacitor undergoing discharge. The standard exponential decay formula would use the initial charge across the capacitor without regard to the history of how the charge was established. The standard model assumes $Q = CV$, so the voltage decays as the charge is dissipated.

Simple first order relaxation in mechanical [20] or electrical systems described here can be written

$$(1 + \tau\, {}_0D_t)\, f(t) = 0. \tag{1.26}$$

where ${}_0D_t$ is the traditional first order derivative, and $f(t)$ represents a physical variable such as electrical polarization or mechanical strain, and τ represents a time scaling constant inherent in the dynamics. This is often written

$$\dot{f}(t) = -\frac{1}{\tau} f(t). \tag{1.27}$$

The formal solution is

$$f(t) = f_0 - \frac{1}{\tau} \int_0^t f(t')\mathrm{d}t'. \tag{1.28}$$

We can generalize the system to allow nonexponential response by replacing the integral operator (the anti-derivative) $(1/\tau)\, {}_0D_t^{-1}$ with $(1/\tau)^\alpha\, {}_0D_t^{-\alpha}$, to obtain

$$f(t) - f_0 = -\frac{1}{\tau^\alpha}\, {}_0D_t^{-\alpha} f(t). \tag{1.29}$$

Applying $\tau^\alpha\, {}_0D_t^\alpha$ from the left and using the differential rule for a constant, ${}_0D_t^{-\alpha} f_0 = f_0 t^{-\alpha}/\Gamma(1-\alpha)$, results in

$${}_0D_t^\alpha f(t) - f_0 \frac{t^{-\alpha}}{\Gamma(1-\alpha)} = -\tau^{-\alpha} f(t). \tag{1.30}$$

From here, we apply the Laplace transform, leading to

$$\tilde{f}(s) = f_0 \frac{s^{-1}}{1 + (s\tau)^{-\alpha}}. \tag{1.31}$$

The series expansion solution is the Mittag-Leffler function.

$$f(t) = f_0 \sum_{k=0}^\infty \frac{(-1)^k}{\Gamma(1+\alpha k)} \left(\frac{t}{\tau}\right)^{\alpha k}, \tag{1.32}$$

from which we recover the decaying exponential in the limit $\alpha \to 1$.

Time domain plots of (1.32) with various values of α are shown in Fig. 1.6. Even for values of the exponent near unity, there is still a transition to power–law behavior after some time. The fact that it looks exponential for early times helps explain why this history effect can be hard to detect for high quality low-loss capacitors.

In actual practice, even this description is too simple. A simulation, and subsequent physical measurements, demonstrating the effect of pre-history is based on the circuit in Fig. 1.7. Initially, switch 1 was set to position 2 and switch 2 in position 3 for an

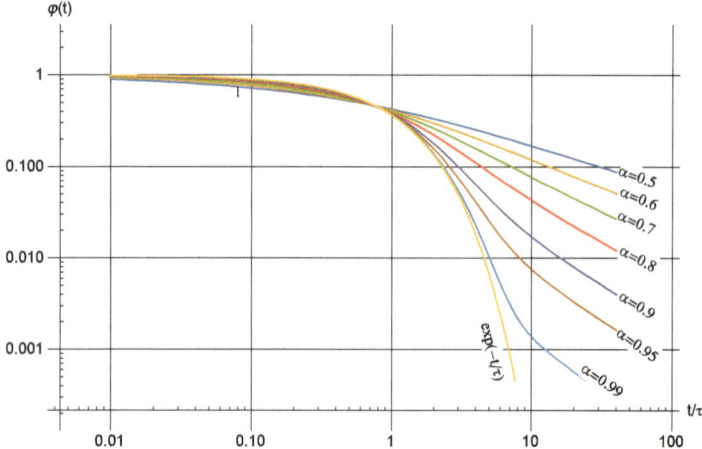

Fig. 1.6 Relaxation plots for various values of the exponent α in (1.32). Exponential decay $e^{-t/\tau}$ is also plotted for comparison

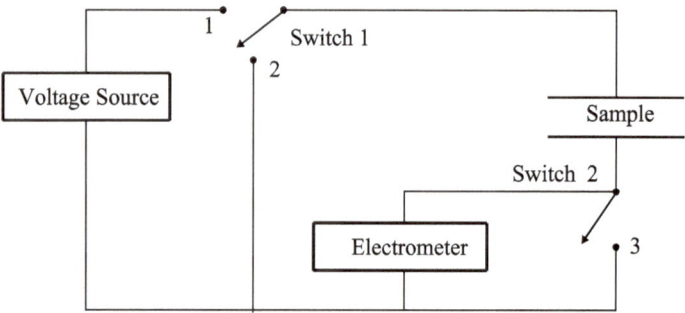

Fig. 1.7 Circuit for recording transient relaxation response

hour or more to ensure that the sample was fully discharged. Switch 1 is then set to position 1 for a specified duration to "charge" the sample. Switch 1 was then set to position 2 to begin discharging the sample. After a couple of seconds to allow for the large surge current to bypass the electrometer, switch 2 was opened to allow recording the decay current. In the simulation here, the sample had an $\alpha = 0.9$ and $\tau = 0.1$ s. Switch 1 was set to charge for 40 τ, then 120 τ, then 600 τ. These pulses generated the pre-history for the current through the sample. Decay curves were then recorded based on the time after switch 1 was set to position 2 (t_0).

In Fig. 1.8, the responses for curves 1–3 were computed based on the entire history of the pulses. For curve 4, only the "initial condition" at t_0 was included. As can be seen the decay histories are quite different. We now understand that the mathematical "initial condition" is for a system that was been in that condition forever prior to the interval of computation. This history effect was confirmed in measurements done at

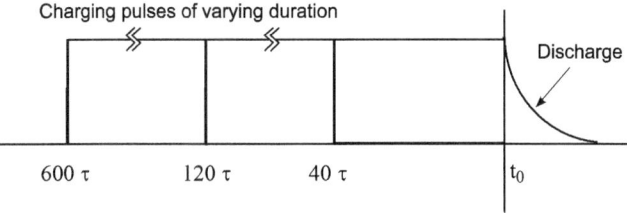

(a) Pre-history of the toy system. Four different "initial" conditions were imposed on the system by charging the system at the same voltage but for different duration.

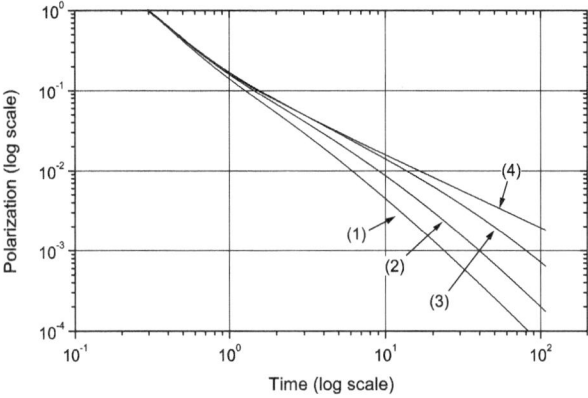

(b) Polarization decay curves resulting from pulses of varying duration.

Fig. 1.8 Polarization decay curves resulting from pulses of varying duration. The toy system has a time constant $\tau = 0.1$ s. Curve (1) is decay after pulse of 40 τ, curve (2) after pulse of 120 τ, curve (3) after pulse of 600 τ. Curve (4) is the fractional relaxation curve predicted by initial conditions

Montana State University—Bozeman in [3]. Other experimental confirmation was included in the text *Fractional Kinetics in Solids*, by Uchaikin and Sibatov [37].

While M. Caputo's attempt to simplify the definition of the FO integral was well intentioned and became popular due to the use of integer order derivatives in the initial condition specification, it is hardly useful in actual practice in time domain analysis. The lesson learned is that the computation must include all of the nonzero history of the input signal. This was made dramatically evident in the article "Dead matter has memory" [42] where Westerlund reported decay currents in a real physical circuit going in the opposite direction from that predicted by the standard techniques. For additional discussion of the initial condition problem see, for example, the article by the CRONE group [27].

Given that our primary interest is in devices with "capacitive like" properties, we see that the FO operators simplify considerably and give the same results, a desirable property when looking for models of real systems. Looking again at the definitions in terms of how they incorporate past history and highlighting the memory kernel in the integral forms and the history weighting in the limit sum form:

$$_aI_t^\alpha f(t) = \int_a^t \underbrace{\frac{(t-t')^{\alpha-1}}{\Gamma(\alpha)}}_{\text{memory kernel}} f(t')\,dt', \quad 0 < \alpha < 1, \tag{1.33}$$

$$_aI_t^\alpha f(t) = \lim_{N\to\infty} \left\{ \left[\frac{t-a}{N}\right]^\alpha \sum_{k=0}^{N-1} \underbrace{\frac{\Gamma(k+\alpha)}{\Gamma(\alpha)\Gamma(k+1)}}_{\text{memory weights}} f\left(t - k\left[\frac{t-a}{N}\right]\right). \right\} \tag{1.34}$$

One would be justified in asking why something so seemingly complicated would be desirable. As will be shown, it is the fact that devices based on fractance properties have this "lossy" memory function that makes them so useful. The decimated memory effect comes from the memory kernel in (1.34) and its memory weights. As time goes on, the weight given to older history of the input signal decays in a symptotic manner. For example, this offers the potential for a control system integrator operator that "unwinds itself" while maintaining a desired phase shift, leading to elimination of the overshoot experienced in many, if not most, control systems.

As we will see in Chap. 4, there are more complex permittivity descriptions than the CC model, a couple of them shown here to compare with the CC model:

$$\text{Cole-Cole (CC)}: \qquad \varepsilon^* - \varepsilon_\infty = \frac{\varepsilon_s - \varepsilon_\infty}{1 + (j\omega\tau)^{1-\alpha}}, \tag{1.35a}$$

$$\text{Cole-Davidson (CD)}: \qquad \varepsilon^* - \varepsilon_\infty = \frac{\varepsilon_s - \varepsilon_\infty}{(1 + j\omega\tau)^{1-\alpha}}, \tag{1.35b}$$

$$\text{Havriliak-Nagami (HN)}: \qquad \varepsilon^* - \varepsilon_\infty = \frac{\varepsilon_s - \varepsilon_\infty}{(1 + (j\omega\tau)^{1-\alpha})^\beta}. \tag{1.35c}$$

Again, each of these was initially developed through curve fitting of the spectroscopic data. Numerical analysis of the dynamics predicted by these models using fractional calculus is discussed in reference [36]. Note that these forms can be written with slight variations, such redefinition of parameters, for example, $1 - \alpha \to \alpha$.

The state of the art is still somewhat chaotic, but that may allow for more innovative approaches. How materials exhibiting CD and HN properties might be put to practical use is open to further research. For now, we will proceed with the simplest type of FO dynamics and construct a new class of electronic device, the "fractor."

1.7   The Push for a New Electronic Device

The idea of using specific impedance properties in a circuit as shown in Fig. 1.9 is nothing new. The generalized gain is

$$G = \frac{V_{out}}{V_{in}} = -\frac{Z_{feedback}}{Z_{in}}.$$ (1.36)

Suppose we start with an impedance described by the constant phase element

$$Z_{CPE} = Z_{Fractance} = \frac{K}{(\tau s)^\alpha},$$ (1.37)

where the parameter K is the reference impedance, in ohm, at the reference frequency ω_C. The time scaling constant $\tau = 1/\omega_C$. The order α is taken from the slope of the impedance magnitude on a log–log plot. It can also be derived from the average phase $\alpha = -\phi/90°$.

As discussed above, the term "fractional capacitor" description of (1.14) is still in common use by engineers today.

$$Z_F = \frac{F}{s^\alpha}$$ (1.38)

with $F = K/\tau^\alpha$ from (1.37) above. Note that as $\alpha \to 1$, F takes on the units of ohm per second. This form highlights the noninteger order properties of capacitors at the expense of clarity as to how one obtains the fractional-order units of the description. It can cause difficulties in designing circuits using FO elements. Nonetheless, this form is accepted as an alternative to (1.37) above and will be used regularly throughout the rest of this brief. In any case, we can now interpret the time domain equation:

$$v(t) = \frac{K}{\tau^\alpha} \, {}_{t_0}I_t^\alpha \, i(t')$$ (1.39a)

or

$$v(t) = K \left[\frac{\Delta t}{\tau}\right]^\alpha \sum_{k=0}^{N-1} \frac{\Gamma(k+\alpha)}{\Gamma(\alpha)\Gamma(k+1)} \, i\,(t - k\Delta t),$$ (1.39b)

Fig. 1.9 Generalized mathematical "operator" using an operational amplifier

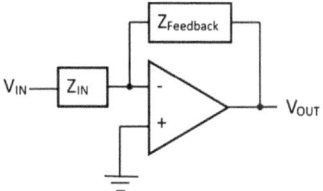

where t_0 is the beginning of the time interval in which current becomes nonzero and $\Delta t = (t - t_0)/N$. The scaling parameter τ now shows up as exactly what is needed to ensure the results appear in real integer order units. As $\alpha \rightarrow 1$, the weights in (1.39b) approach 1 for all k and the integer order integral is obtained. In other words, the ideal capacitor accumulates perfect memory of the net integrated current.

If we put a fractance in the feedback position and let the input impedance to the opamp be a standard resistor $Z_{in} \rightarrow R_{in}$, then $i_{in}(t) = v_{in}(t)R_{in}$. This allows us to write the output voltage in terms of the input voltage

$$v_{out}(t) = \frac{K}{R_{in}} \frac{1}{\tau^\alpha} \, {}_{t_0}I_t^\alpha v_{in}(t'), \tag{1.40}$$

and we have an analog FO integrator. Again, the scaling constant τ^α ensures that the resulting units are consistent. The epiphany here is that this is nothing more than a generalization of the idea of making an integrator from a capacitor and an operational amplifier. This kind of circuit was suggested as early as 1961 in [7], but now possible with a single feedback element rather than a ladder constructed of several resistors and capacitors.

The FO integrator was tested with one of the first solid-state fractors with an order $\alpha \approx 0.5$ discussed in the next chapter. The $^1/_2$ order integral of a unit step function goes as $\sqrt{t}$. Figure 1.10 exhibits this response.

The ideal fractance device would isolate the power–law CPE behavior and produce something as close as possible to the plot shown in Fig. 1.1. We can now think of the ideal capacitor as a fractance with α approaching unity, but other "ideal" elements could be formed with arbitrary values of α. How close we get to the ideal will be part of the discussion of Chap. 2. As we have been developing experience working with prototypes of fractance devices, we note that higher frequency phase variation of as much as $\phi \approx \pm 5°$ does not adversely affect the accuracy of the operation. In other words, pushing for the "ideal" fractance behavior may lead to diminishing returns. What we need is a set of fractances with a variety of fraction dynamic orders, covering a broad range of frequency bands to offer these devices to the widest range

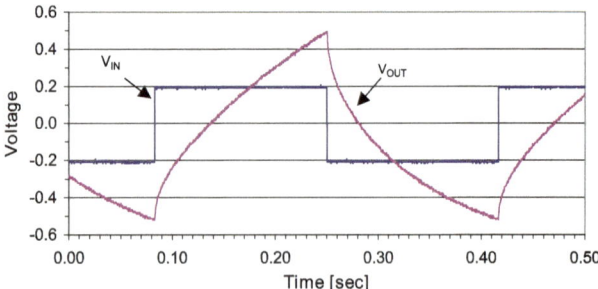

Fig. 1.10 Input signal was a *square wave*. The output clearly exhibits the square root of time behavior. From [5]

of applications. While we discuss progress on achieving this goal of implementing fractance properties in a single analog element, it must be realized that a set of such fractance devices, with varying reference impedance magnitudes and a variety of fractional-orders will be required to cover slow mechanical control systems up through high speed signal processing.

References

1. E. Barsoukov, J.R. Macdonald, *Impedance Spectroscopy: Theory, Experiment and Applications*, 2nd edn. (Wiley, Hoboken, New Jersey, 2005)
2. H. Bode, *Network Analysis and Feedback Amplier Design* (Van Nostrand, New York, 1945)
3. G.W. Bohannan, in Application of fractional calculus to polarization dynamics in solid dielectric materials. Ph.D. Thesis, Montana State University—Bozeman (2000)
4. G.W. Bohannan, Analog realization of a fractional controller, revisited, in *Tutorial Workshop 2: Fractional Calculus Applications in Automatic Control and Robotics*, Las Vegas, USA, ed. by B.M. Vinagre, Y.Q. Chen (2002), pp. 175–182
5. G.W. Bohannan, Analog fractional order controller in temperature and motor control applications. J. Vibr. Control **14**(9–10), 1487–1498 (2008)
6. R. Caponetto, D. Porto, Analog implementation of non integer order integrator via eld programmable analog array, in *FDA'06: Proceedings of the 2nd IFAC Workshop on Fractional Differentiation and its Applications*, Porto, Portugal (2006), pp. 170–173
7. G.E. Carlson, C.A. Halijak, Simulation of the fractional derivative operator $\sqrt{s}$ and the fractional integral operator $1/\sqrt{s}$, in *Central States Simulation Council Meeting on Extrapolation of Analog Computation Methods*, Kansas State University, vol. 45, no. 7 (1961), pp. 1–22
8. Y.-Q. Chen, K.L. Moore, Discretization schemes for fractional–order differentiators and integrators. IEEE Trans. Circuits Syst.–I: Fund. Theory Appl. **49**(3), 363–367 (2002)
9. Y.-Q. Chen, B.M. Vinagre, I. Podlubny, Continued fraction expansion approaches to discretizing fractional order derivatives-an expository review. Nonlinear Dyn. **38**, 155–170 (2004)
10. Y.Q. Chen, Tuning methods for fractional–order controllers, U.S. Patent 7,599,752
11. K.S. Cole, R.H. Cole, Dispersion and absorption in dielectrics, J. Chem. Phys. **9**, 341–351 (1941)
12. P. Debye, *Polar Molecules* (Chemical Catalogue Company, New York, 1929)
13. W. Feller, *An Introduction to Probability Theory and Its Applications*, vol. II (Wiley, New York, 1966)
14. M. Filoche, M. Sapoval, Transfer across random versus deterministic interfaces. Phys. Rev. Let. **84**(25), 5776–5779 (2000)
15. P. Grigolini, A. Rocco, B.J. West, Fractional calculus as a macroscopic manifestation of randomness. Phys. Rev. E. **59**(3), 2603–2613 (1999)
16. A. Janicki, A. Weron, *Simulation and Chaotic Behavior of α-Stable Stochastic Processes* (Dekker, New York, 1994)
17. A.K. Jonscher, The "universal" dielectric response. Nature **267**, 673–679 (1977)
18. A.K. Jonscher, *Dielectric Relaxation in Solids* (Chelsea Dielectric Press, London, 1983)
19. S. Manabe, The non-integer integral and its application to control systems. Jpn. Inst. Electr. Eng. J. **80**(860), 589–597 (1960)
20. T. Nonnenmacher, W. Glöckle, A fractional model for mechanical stress relaxation. Phil. Mag. Lett. **64**(2), 89–93 (1991)
21. K. Oldham, J. Spanier, *The Fractional Calculus: Theory and Applications of Differentiation and Integration to Arbitrary Order* (Academic Press, New York, 1974)
22. A. Oustaloup, B. Mathieu, P. Lannusse, The CRONE control of resonant plants: application to a fexible transmission. Eur. J. Control **1**(2) (1995)

23. A. Oustaloup, P. Lanusse, P. Melchior, X. Moreau, J. Sabatier, J.L. Thomas, A survey of the CRONE approach, in *Conference Proceedings 1st IFAC Workshop on Fractional Differentiation and its Applications FDA04*, (2 part) (2004)
24. C.M.A. Pinto, A.M. Lopes, J.A. Tenreiro Machado, A review of power laws in real life phenomena, Commun. Nonlinear Sci. Numer. Simulat. **17**(9), 3558–3578 (2012)
25. I. Podlubny, *Fractional Differential Equations: An Introduction to Fractional Derivatives, Fractional Differential Equations, to Methods of their Solution and Some of their Applications, Mathematics in Science and Engineering*, vol. 198 (Academic Press, San Diego, CA, 1999)
26. I. Podlubny, I. Petráš, B.M. Vinagre, P. O'Leary, L. Dorčák, Analog realizations of fractional-order controllers, Nonlinear Dyn. **29**, 281–296 (2002)
27. J. Sabatier, M. Merveillaut, R. Malti, A. Oustaloup, How to impose physically coherent initial conditions to a fractional system? Commun. Nonlinear Sci. Numer. Simul. **15**, 1318–1326 (2010)
28. J. Sakurai, *Modern Quantum Mechanics*, Revised edn. (Supplement II, Addison-Wessley, Reading, PA, 1994)
29. H.V. Schmidt, J.E. Drumheller, Dielectric properties of lithium hydrazinium sulfate. Phys. Rev. B. **4**(12), 4582–4597 (1971)
30. J.A. Tenreiro Machado, Theory of fractional integrals and derivatives: application to motion control, in *ICRAM95—IEEE/IFAC/ASME/JSME International Conference on Recent Advances in Mechatronics*, 14–16 Aug 1995, Istanbul, Turkey (1995), pp. 1086–1091
31. J.A. Tenreiro Machado, Analysis and design of fractional-order digital control systems, Syst. Anal. Model. Simul. **27**(2–3), 107–122 (1997)
32. J.A. Tenreiro Machado, Fractional-order derivative approximations in discrete-time control systems, Syst. Anal. Model. Simul. **34**, 419–434 (1999)
33. J.A. Tenreiro Machado, V. Kiryakova, F. Mainardi, Recent history of fractional calculus. Commun. Nonlinear Sci. Numer. Simul. Elsevier, **16**(3), 1140–1153 (2011)
34. J.A. Tenreiro Machado, *Shannon Information and Power Law Analysis of the Chromosome Code, Abstract and Applied Analysis*, Hindawi, vol. 2012, Article ID 439089, (13 pp.) (2012)
35. J.A. Tenreiro Machado, C.M.A. Pinto, A.M. Lopes, A review on the characterization of signals and systems by power law distributions. Signal Process. **107**, 246–253 (2015)
36. J.A. Tenreiro Machado, Matrix fractional systems. Commun. Nonlinear Sci. Numer. Simulat. **25**, 1018 (2015)
37. V. Uchaikin, R. Sibatov, *Fractional Kinetics in Solids: Anomalous Charge Transport in Semiconductors* (Dielectrics and Nanosystems, World Scientific, Singapore, 2013)
38. D. Valério, J.A. Tenreiro Machado, V. Kiryakova, Some pioneers of the applications of fractional calculus. Fractional Calc. Appl. Anal. **17**(2), 552–578 (2014). doi:10.2478/s13540-014-0185-1
39. E. Warburg, Uber das Verhalten sogenannter unpolarisierbarer Electroden gegen Wechselstrom. Ann. Phys. Chem. **67**, 493–499 (1899)
40. B.J. West, *Physiology, Promiscuity and Prophecy at the Millennium: A Tale of Tails* (World Scientific, Singapore, 1999)
41. B.J. West, *Fractional Calculus View of Complexity: Tomorrow's Science* (CRC Press, Boca Raton, 2016)
42. S. Westerlund, Dead matter has memory. Phys. Scr. **43**, 174–179 (1991)
43. S. Westerlund, L. Ekstam, Capacitor theory. IEEE Trans. Dielectr. Electr. Insul. **1**(5), 826–839 (1994)

Chapter 2
Devices

2.1 Introduction

In this chapter, we introduce some fractional-order devices (FOD). The successful recipes allowed for the construction of prototype devices and demonstrated the ease of designing circuits and systems using these devices. Some demonstrations of these devices in circuits are included in Chap. 3. No attempt will be made to discuss all possible approaches for fabricating such devices. The goal is to point the way to creating devices with stable, specific phase over broad frequency ranges.

To be specific, we are not looking at "fractal capacitors," such as described in [40] where the area of the capacitor plates is increased by the use of fractal geometry. That technique makes better use of available surface area to achieve higher capacitance values. The devices discussed here exhibit "fractional-order dynamics" with both energy storage and dissipation properties being described by an impedance of non-integer order.

Electrical impedance spectroscopy (EIS) techniques have been steadily developing over the last century. By the middle of the twentieth century, the description of a constant phase element (CPE) had been invented [19]. The implications of this description were not readily apparent early on. Subsequent study suggested that analysis of CPE aspects might be useful for diagnostic purposes, such as corrosion detection, for example. As discussed in Chap. 1, by the turn of the millennium it was suggested that a material exhibiting CPE behavior in a might offer an approach to creating a new class of circuit elements.

We make the distinction between the CPE properties of materials forming fractional-order elements (FOE) and the ultimate fractional-order devices (FOD) made using these materials.

The reader should note that there are slight variations in the terminology used for the various FOE and FOD described herein. The authors came together to present their efforts from their perspectives. The state of the art is not yet ready for full standardization. The process of creating a new class of electronic device has been somewhat chaotic and has required each of us to consider many different approaches. Our overall goal is to pass on these perspectives and stimulate new approaches.

© The Author(s) 2017   21
R. Caponetto et al., *Fractional-Order Devices*,
SpringerBriefs in Nonlinear Circuits, DOI 10.1007/978-3-319-54460-1_2

2.2 Discrete Element Approximations of Fractional-Order Elements

Numerous attempts have been made to approximate fractance response [10, 37, 44, 48], as well as multiple attempts at creating digital operators for control and signal processing [12, 14, 16, 17, 32, 34, 36, 49]. These approximations all are based on one or another self-similar arrangements, sometimes referred to as deterministic fractal scaling. A popular form is the continued fraction expansion,

$$Z(s) = R_1 + \cfrac{1}{C_1 s + \cfrac{1}{R_2 + \cfrac{1}{C_2 s + \cdots}}}, \tag{2.1}$$

with $R_{n+1} = R_n \cdot R_{scale}$ and $C_{n+1} = C_n \cdot C_{scale}$, with R_{scale} and C_{scale} chosen to form a self-similar power-law spacing of the time constants. The ratio of R_{scale} to C_{scale} determines the phase angle. The values of R_1 and C_1 determine the upper frequency for the CPE range. For a given number of circuit elements, the closer R_{scale} and C_{scale} are to unity, the less the phase ripple, but with reduced bandwidth. The greater R_{scale} and C_{scale}, the broader the bandwidth, but greater phase ripple.

For example, an approximation of a $-35°$ fractance circuit is shown in Fig. 2.1 and its impedance spectrum is shown in Fig. 2.2. Using such a circuit in practice is problematic for several reasons. Not the least of which is that the values of the elements given in the design are typically not readily available and often need to be built with parallel/series equivalent circuits, making the device even more complicated, bulky, and expensive.

Approximated FOD have been fabricated based on the self-similar scaling rules and exhibit the phase ripple as suggested in the example above [25]. Impedance spectroscopy texts offer a range of self-similar scaled equivalent circuit topologies that would produce CPE behavior in order to describe the observed data in disordered media [5].

The implication from this exercise is that if we could create a very broad distribution of time constants with a vast number of microscopic elements then we

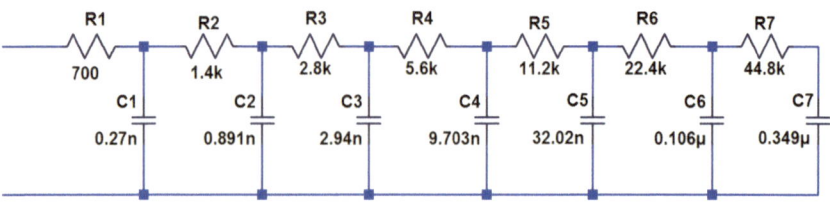

Fig. 2.1 Approximation of a $-35°$ phase fractance using standard discrete R and C elements. For the circuit shown here: $R_1 = 700\ \Omega$, $C_1 = 0.27$ nF, $R_{scale} = 2.0$, $C_{scale} = 3.3$

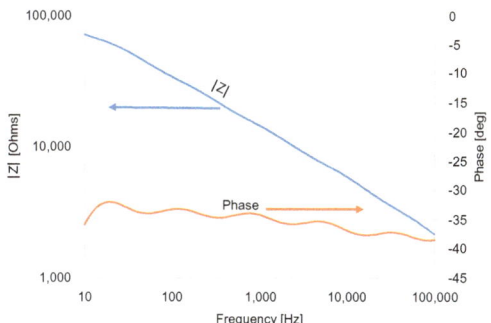

Fig. 2.2 Phase and magnitude for a fractance approximation using standard discrete R and C elements. For the plot shown here: $R_1 = 700\ \Omega$, $C_1 = 0.27$ nF, $R_{scale} = 2.0$, $C_{scale} = 3.3$. Note the phase ripple over the frequency band of $\Delta\phi \approx < \pm 5°$

could achieve low-phase ripple and wide effective bandwidth, if the natural distribution turns out to follow self-similar scaling. Fortunately, this self-similar scaling or random fractal scaling seems to be ubiquitous. The challenge is isolating the CPE behavior from the other processes.

A device based on deterministic geometric fractal patterns on a silicon chip had been introduced by Haba et al. in [25], but the technique used applied only to high-frequency devices ($f > 100$ kHz) which would be appropriate for many signal processing applications but not to low-frequency motor control problems. To extend this design technique down to mHz would require many square meters of substrate. A different approach would be required; in this case, making use of intrinsic properties of disordered dielectric materials.

2.3 Early Fractional-Order Devices (FOD)

There have been several generations of attempts at creating single-element FOD. They virtually all start with the idea of creating some form of anomalous charge transport behavior. Free charges can flow in a material due to drift induced by an electric field or by concentration gradients where like charges repel each other resulting in diffusion. In either case, the sum of all the motions contributes to the net current. Randomizing the conduction and/or diffusion paths then creates the distribution of flow statistics. Many different conceptual models have been proposed, such as random walks with a distribution of waiting times due to combination of the charge carrier with a fixed ion then regenerating the mobile carrier.

2.3.1 Platinum Nanowire in Polymer

The initial attempt at creating a fractance was based on growing cross-linked conductive threads from conductive $K_2[Pt(CN)_4Br_{0.3}]$ crystals in a nonconductive liquid polymer solution. After drying, the polymer film was cut and mounted against

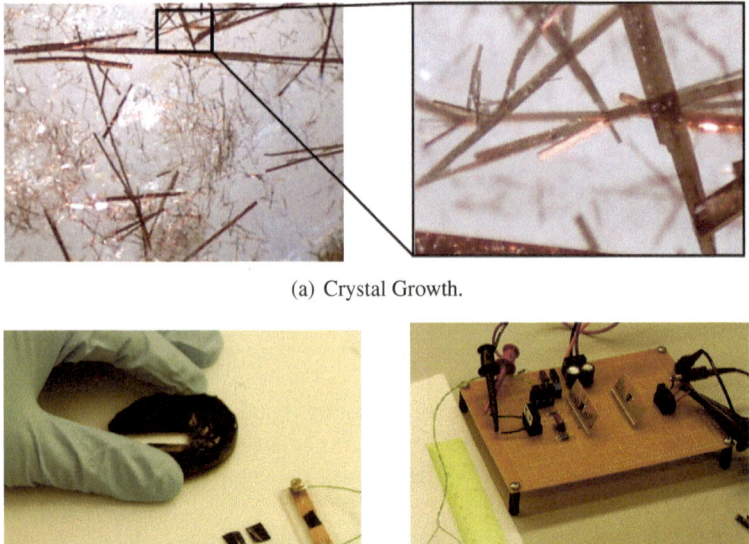

(a) Crystal Growth.

(b) Cut for testing.   (c) Mounted for impedance measure-
                                         ment.

Fig. 2.3 Growing and testing Pt chains in a polymer

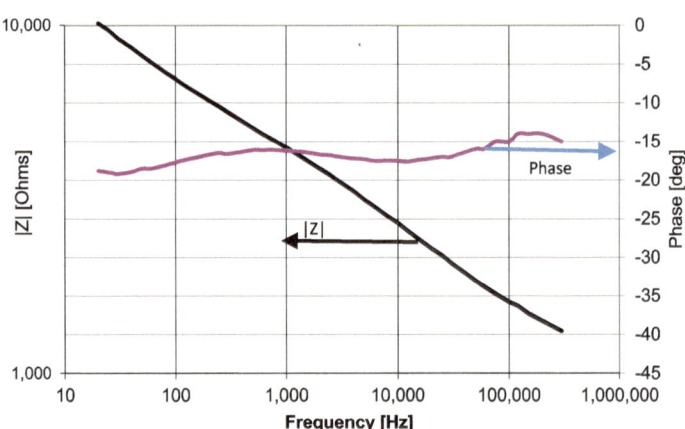

Fig. 2.4 Impedance spectrum for a sample of the $K_2[Pt(CN)_4Br_{0.3}]$ crystals in polymer film

electrodes for impedance testing. Randomized conductive paths interspersed with
capacitive cross-links did induce a CPE type of response. This first attempt showed
that there was promise in isolating near constant phase over a fairly broad frequency
range. This process showed promise, but it was expensive and the yield of samples
showing fractance characteristics was less that 25% (Figs. 2.3 and 2.4).

2.3.2 Lithium-Ion Type

The next approach was to make use of anomalous diffusion rather than direct charge drift under the influence of an electric field while making use of the success of randomly cross-linked chains. This minimized the effect of the blocking and parasitic capacitances in Fig. 1.5.

In order to get the very broad frequency response desired, we looked at randomizing the structure of the charge transport pathways in the substrate. This was achieved in two steps. The randomization of the charge flow paths is enhanced by first roughening the inner plate, then dipping it into a suspension of the $LiNO_4$, doped polyananline in sulfonic acid and tetraethyl orthosilicate (TEOS). The roughening allowed the long polyananline strands to twist around in random patterns. Each crossing of one strand to another generated a trapping/regeneration site. Each dipping was oriented 90° from the previous dipping. This further enhanced the randomization of the charge paths. The silicate gel was to stabilize the structure upon drying [35].

Figure 2.5a, starting at the upper right, shows a clean copper plate cut to ≈2.5 cm on a side. At the bottom is a roughened plate. At the top left is a plate after coating. After each dipping, the plates were dried for 30 to 50 min in a low temperature oven at ≈90 °C. Higher temperatures tend to cause the plating to flake off. The rack is allowed for processing multiple samples at once. Twelve or more dipping and drying cycles

(a) Inner plate preparation.

(b) Plate Dipping.

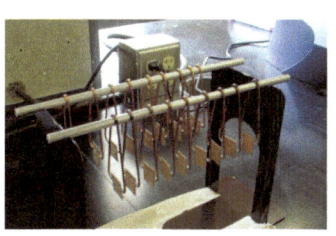

(c) The drying rack.

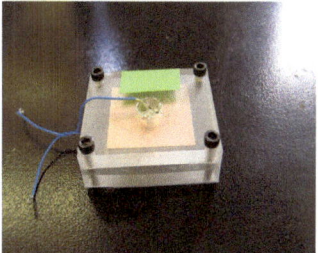

(d) Assembled Prototype.

Fig. 2.5 Fractor preparation process. The coated inner plate is mounted between two outer electrode plates. The copper plates act as insulators for the lithium ions and the nitrate radicals. The mobile electrons in the organic chains are bound to the chains. With this, there is no direct mobile charge flow from the *upper* to the *lower* electrode plates

Fig. 2.6 An edge view of the solid-state fractor assembly showing the current paths in the electrolyte layer. By having the charge carriers move orthogonally to the applied field, they exhibit diffusive charge transport as opposed to drift which they would exhibit if they moved directly with the applied field

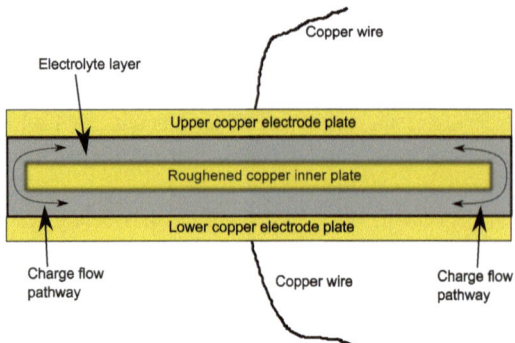

were required to achieve the desired response. The coated plates were sandwiched between two outer electrode plates and the assembled device coated in a plastic dip to seal it from external moisture. The final product was dubbed the "fractor."

In Fig. 2.6, the idea is to have the bulk of the charge carriers diffuse orthogonally to the applied field. Charges on edge will flow with the field around the central layer. It is this edge current that is seen as the external circuit. Because the charge becomes deleted or enhanced on the edges, the external circuit sees a decreasing or increasing current for a constant imposed potential. Charge diffuses out to the edges due to concentration gradients, not directly due to the applied field. This translates the anomalous diffusion statistics into a sum distribution in the current in the external circuit.

The randomization of the pathways gives rise to a broad distribution of "time constants" for trapping and recombination–regeneration of the carriers. The model of current flow (assumed to be Li^+ ions) around the edges of the inner plate was confirmed by grinding off the electrolyte around all the four edges, with only bare copper exposed. The resulting impedance was almost purely capacitive ($\alpha \approx 1$), as if the electrolyte had become a high-quality insulating dielectric. A typical impedance spectrum is shown in Fig. 2.7.

Unfortunately, these devices exhibited slow degradation over time, but not so quickly as to prohibit testing[1] of these devices in physical circuits in a variety of applications to be discussed in Chap. 3. This will require more work on the recipes for fabricating the devices. Organic polymers are subject to deterioration and may be one contribution to the aging. Over the course of one to four years, the conductivity decreased substantially, with the impedance magnitude increasing by factors of 10^2 to 10^3, as shown in Fig. 2.8. Interestingly, the phase did not seem to change, with the implication that the randomization statistics do not change over time. The most likely cause of the degradation is loss of Li^+ ion concentration due to interaction with oxygen forming an immobile Li_2O. The roll-off at the higher frequencies due to the parasitic capacitance of the copper plates becomes even more apparent.

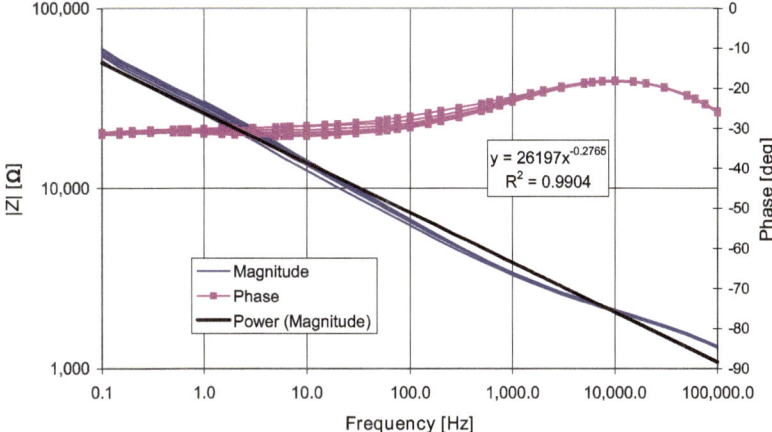

Fig. 2.7 Impedance plot. The phase roll-off at the *upper* frequencies is due to the parasitic capacitance of the electrodes. Thickening of the electrolyte layers pushes the roll-off to higher frequencies

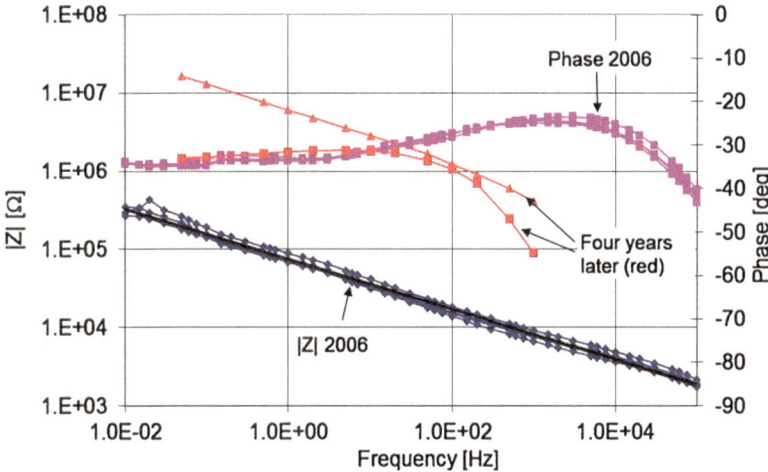

Fig. 2.8 Typical aging of solid-state fractor properties over a 4-year period. Impedance spectroscopy results of a fractor manufactured in 2006. The *pink line* represents the phase angle measurements while the *blue line* measures the impedance. The *red line* composed of triangles represents the impedance, where the *red-square line* represents the phase angle measurements, of the same fractor manufactured in 2006 measured again in 2010. The roll-off at the higher frequencies is due to the parasitic capacitance of the outer copper electrode plates

From there a manufacturing process will need to be invented that allows for production of predictable and consistent K and α values, with these values selectable on a batch-by-batch basis.

2.3.3 Modified Lithium-Ion Type

This second recipe was necessitated by the discontinued production of the polyanaline in sulfonic acid. Attempts to synthesize polyanaline in sulfonic acid locally proved the futile.

A revised recipe with cuprous mercuric iodide formed from potassium iodide, copper sulfate, mercuric nitrate, acetic acid, and acetone was tried with the intent to form a more consistent porous ionic gel. This proved to be very temperature dependent and was abandoned. The next step was to just to try the earlier recipe without the polyanaline (Fig. 2.9).

The materials currently in use are lithium nitrate ($LiNO_3$), tetraethyl orthosilicate (TEOS, $Si(C_2H_5)_4$), distilled water (H_2O), 95% ethyl alcohol (C_2H_5OH), and nitric acid (HNO_3).

Ionic Gel Preparation: 7 mL of 95% ethyl alcohol was mixed with 7 mL of distilled water in a beaker with a magnetic stir rod and 0.7 g of lithium nitrate was added to it. The beaker was mounted on a stir plate and stirred the mixture until the entire lithium nitrate is solvated and removed the solution from the stir plate. Next, 3.5 mL of TEOS was added to the solution, at first the TEOS is not be soluble in the solution, so a drop or two of nitric acid must be added to the mixture. The beaker containing the solution was covered with a Petri dish to reduce degradation and evaporation and mounted back onto the stir plate and stirred on to a setting of approximately 300 rpm and left to stir for approximately 2 h. The beaker was tightly covered for storage. The solution must be used quickly as the compound degrades rapidly.

Again, we coated roughened square copper plates, approximately 3 mm thick, 2.5 cm on a side. The copper plates are dip coated and are held by their edges using bent copper wiring ("tweezer clamps") and dipped in the solution, and then hung to dry on a rack. Once the coating has dried (approximately 5–15 min depending on how many coatings have already been applied), the copper plates were then removed from the clamps, rotated by 90° to ensure the sides of the plates are all coated, and placed back into the clamps, dipped and hung to dry again. This process was repeated to build up a minimum of 14 dip coatings on the copper plates in order to build appropriate ionic gel thickness.

Additional work needs to be done to determine the precise recipe variations lead to specific fractional-orders. Clues for varying the order can be developed from the IPMC, CBNC, and solution-based systems to be discussed in the next sections. Other clues could be taken from [22] in which it is shown that ionic conductivity is dependent on the concentration of the lithium ions and on the structure of the polymer electrolyte. Since fractors have been demonstrated with orders ranging between 0.27 and 0.5, it is possible that we have seen this concentration difference effect already merely in slight differences on a batch-to-batch basis.

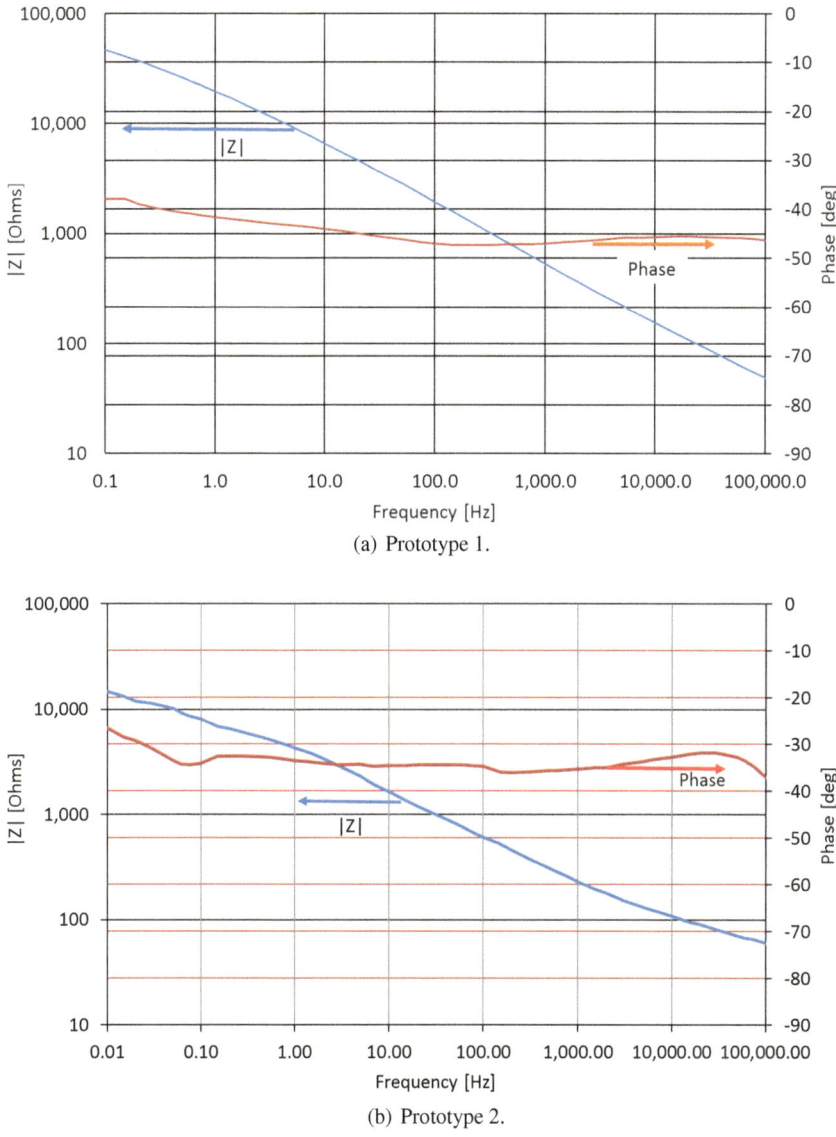

(a) Prototype 1.

(b) Prototype 2.

Fig. 2.9 Impedance spectra for the second-generation fractor prototypes used later in the fractional harmonic oscillator demonstration

2.3.4 Other Lithium-Ion Attempts

We tried an idea for very low-cost fractance by just doping standard gelatin with lithium nitrate. There was evidence for CPE behavior, but the distribution of time constants was not adequate to get broad band phase consistency (Fig. 2.10).

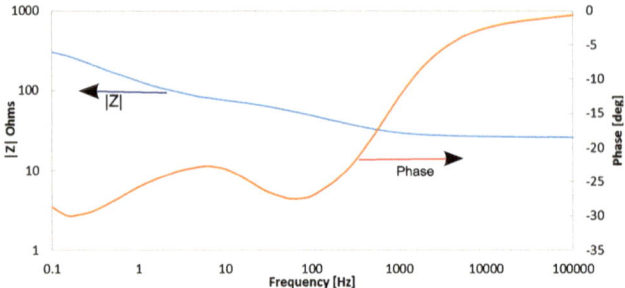

Fig. 2.10 An attempt using off-the-shelf gelatin doped with lithium nitrate exhibiting a mixture of CPE properties

2.4   Nanostructured Materials as Fractional-Order Elements

Ionic polymeric–metal composite (IPMC) and carbon-based nanocomposite (CBNC) are innovative nanostructured materials. The aim of this section is to show the fractional properties of IPMC and CBNC.

By varying some realization parameters, both for the IPMC and the CNNC, such as the geometrical dimensions, the duration of the electrodes deposition, the cation type, the solvent, the carbon percentage, and the curing temperature, it seems possible to obtain models, integrators, and derivators, whose non-integer order is strictly related to such parameters.

The interest of current investigation in fact is focused on the possibility to use IPMC and CBNC to realize fractional-order elements (FOE) able to implement a fractional-order device.

2.4.1   IPMC Structure and Working Principles

Ionic polymer–metal composites (IPMCs) as electroactive polymers (EAPs) have the very useful capabilities to transform electrical energy into mechanical energy, and vice versa [41, 42], making them privileged candidates for the implementation of actuators or sensors with features as low applied voltage, high compliance, softness, etc., thus creating great interest in future applications in very different fields such as aerospace, biomedics, and robotics [18, 38, 43].

In [15], the possibility of modeling the IPMC actuators via gray-box model based on fractional-order systems was presented, paving the path for a new approach to such materials seen as fractional-order electronic elements (FOEs) and not only as electromechanical transducers.

This new approach is suggested by IPMC electrochemical and structural properties. In fact the dendrites on the interfacial landscape between the metal electrodes

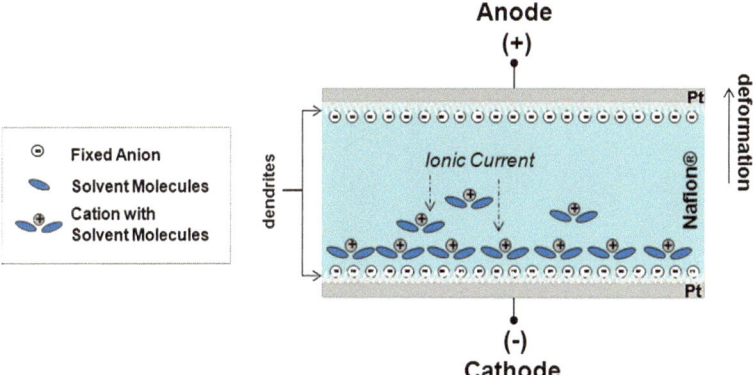

Fig. 2.11 Structure and working principle of IPMC

and the polymer layer show fractal dimensions, and a fractional electrical behavior might be due to the anomalous diffusion of ions and solvent through metal/polymer surface, see [28]. Furthermore, IPMCs fit in the research area working on FOE implementation processes which assesses the interaction between ionic phenomena and fractal structures as possible starting point for FOEs' realization.

IPMCs are based on polymer-containing ions, ionomers, or ionic polymers, that are weakly linked to the polymer chain and metallized via a chemical process, on both sides, with a noble metal, to realize the electrodes. There are a number of different types of ionic polymers available but the typical IPMC used in many investigations is composed of a perfluorinated ion-exchange membrane, Nafion 117, which is surface-composited by platinum via chemical process, see Fig. 2.11.

The platinum electrodes usually consist of small, interconnected particles of metal which are made to penetrate into the ionic polymer membrane. This allows the formation of electrodes with dendritic structures [28, 41], which extend from the surface into the membrane.

Working as an actuator, when an external voltage is applied across the thickness of the IPMC, mobile cations H^+ in the polymer will move toward the cathode. If a solvent is present in the sample, the cations will carry solvent molecules with them. The cathode area will expand while the anode area will shrink. If the tip of the IPMC strip is free the polymer will bend toward the anode; thus a force will be delivered. On the other hand, when the IPMC works as a sensor, it exploits the mechanical displacement of the polymer for the generation of a ionic current inducing a potential difference.

2.4.1.1 Manufacturing

Nafion 117 film, by DuPontTM, Sigma-Aldrich Group, with thickness $t_{Naf} = 180\,\mu m$, and sizes 4 cm × 4 cm, was pretreated by successive boiling for 30 min in HCl_2N and deionized water.

Fig. 2.12 IPMC sample

Ethylene glycol (EG) was used as the solvents and Platinum as the electrodes. Two platinum metallizations were obtained after the immersion of the Nafion117 membrane in a solution of $[P_t(NH_3)_4]Cl_2$, with MW = 334.12, and immersion time will be here referred to as absorption time.

The platinum solution was obtained by dissolving 205 mg of the complex in 60 ml of deionized water and adding 1 ml of ammonium hydroxide at 5%.

In order to increase the performance of the device, a dispersing agent (polyvinyl pyrrolidone with molecular weight 10000-PVP10) has been added. A secondary metallization was performed via deposition. Then, the samples were boiled in 0.1 M HCl for 1 h.

In order to obtain the IPMC with EG as solvent, Nafion117 membranes were soaked overnight in a beaker containing pure EG and, finally, heated to 60 °C for 1 h. Obtained IPMC was then cut into strips of size 1 cm × 1 cm and dried for one week. Some IPMC samples are shown in Fig. 2.12.

Three different IPMC membranes have been fabricated with three different absorption times: 5, 10, and 20 h in order to study the relationship between such a fabrication parameter and the fractional-order dynamics of the IPMC device.

The three membranes will be here referred, respectively, as $IPMC_{AbsT-5h}$, $IPMC_{AbsT-10h}$, and $IPMC_{AbsT-20h}$.

2.4.1.2 Geometry and Experimental Setup

In the experiments the IPMC device proposed as FOE consists of an IPMC sample (1 cm × 1 cm) mechanically fixed within a plexiglas sandwich configuration in series with a resistor with $R = 46\ \Omega$ as shown in the schematic of Fig. 2.13.

Fig. 2.13 IPMC FOE sandwich configuration schematic

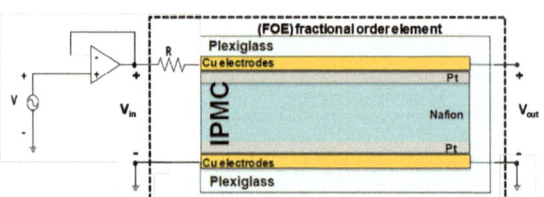

Fig. 2.14 Experimental setup (*left*), IPMC FOE setup (*right*)

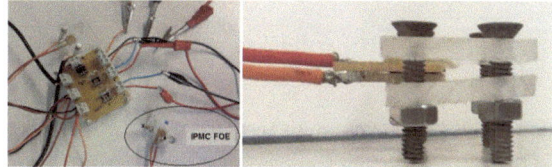

The input voltage signal V_{in}, applied to perform the frequency-domain identification, was forced by a waveform generator Agilent 33220A, through a conditioning circuit made of an operational amplifies, ST TL082CP, in buffer configuration. The output voltage V_{out} was measured through a pair of electrodes 1 cm × 1 cm and thickness 35 μm printed on a PBC board and in direct contact with the platinum electrodes.

Both the input V_{in} and output V_{out} signals were acquired using a National Instrument NI USB-6251 board and processed by the LabView software.

A schematic of the experimental setup is reported in left part of Fig. 2.14, while details on the IPMC FOE are in right part of Fig. 2.14.

Measurements have been performed on the three membranes $IPMC_{AbsT-5h}$, $IPMC_{AbsT-10h}$, and $IPMC_{AbsT-20h}$. A set of sine voltages with amplitude of 4 V_{pp} was applied as V_{in}.

For each membrane, different measures were performed varying V_{in} frequency in the range from 10 mHz to 10 kHz with 10 Hz step. MATLAB tools were used to estimate the modulus and phase of the acquired signals.

2.4.2 Linearity Study

As reported in literature [11], IPMC membranes working as transducers present a nonlinear component in the electromechanical model and show an hysteretic behavior in the relationship between applied voltage and absorbed current, and therefore the hypothesis of linearity must be verified in order to consider the frequency response as a coherent characterization for IPMC. Moreover, being the system nonlinear, such characterization is valid only for the given input voltage amplitude 4 V_{pp}.

Lissajous curves have been studied in order to characterize the linearity of the IPMC as FOE. A Lissajous curve was obtained at each frequency for the entire experimental range. Figures 2.15, 2.16, 2.17, and 2.18 show such curves for the device $IPMC_{AbsT-5h}$. Moreover, Fig. 2.19 shows zoomed curves for the sample frequencies: $f = 50$ mHz, $f = 1$ Hz, $f = 9$ Hz, and $f = 4$ kHz.

At low frequencies, the nonlinear component dominates and the Lissajous curves have a non-elliptic shape. In particular under the 1 Hz frequency, the curves show nonlinearity, while 1 Hz is a frequency of transition from nonlinearity to linearity. For frequencies higher than 1 Hz the shape of the Lissajous curves can be considered elliptic.

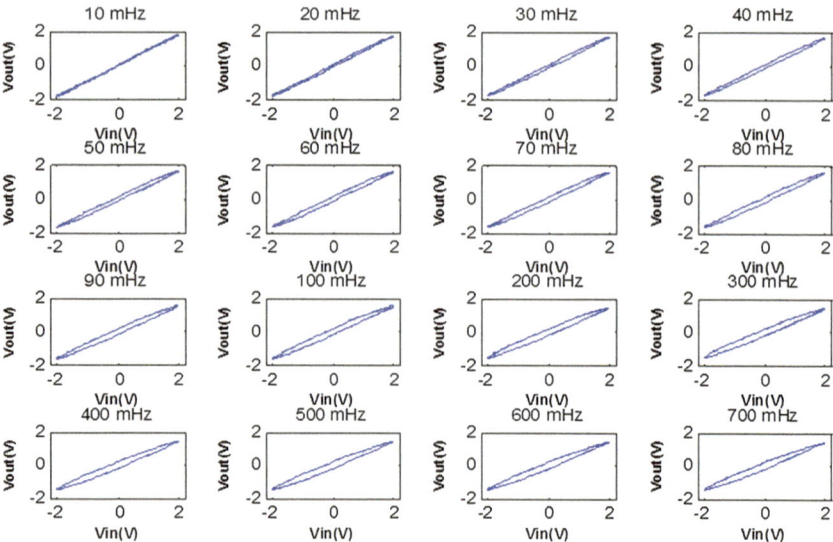

Fig. 2.15 Lissajous curves on measured input V_{in} and output V_{out} voltages in the frequency range between 10 and 700 mHz for the device IPMC$_{AbsT-5h}$

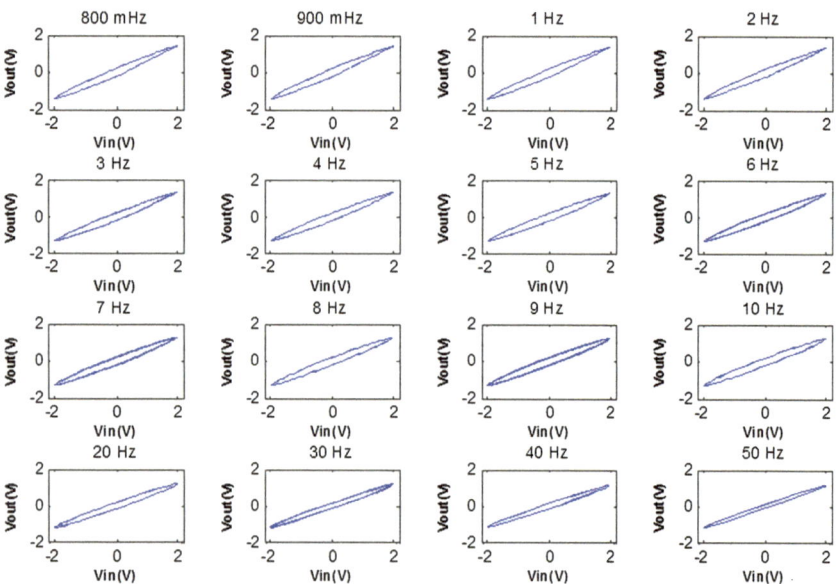

Fig. 2.16 Lissajous curves on measured input V_{in} and output V_{out} voltages in the frequency range between 800 and 50 Hz for the device IPMC$_{AbsT-5h}$

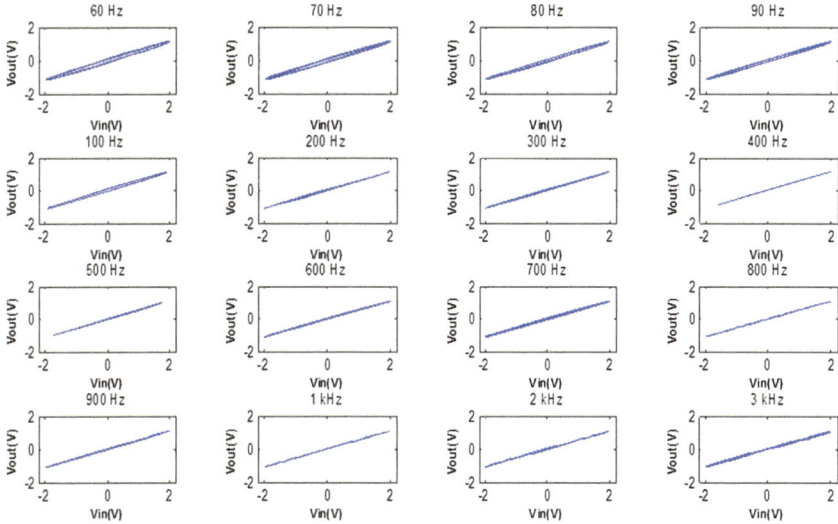

Fig. 2.17 Lissajous curves on measured input V_{in} and output V_{in} voltages in the frequency range between 60 Hz and 3 kHz for the device IPMC$_{AbsT-5h}$

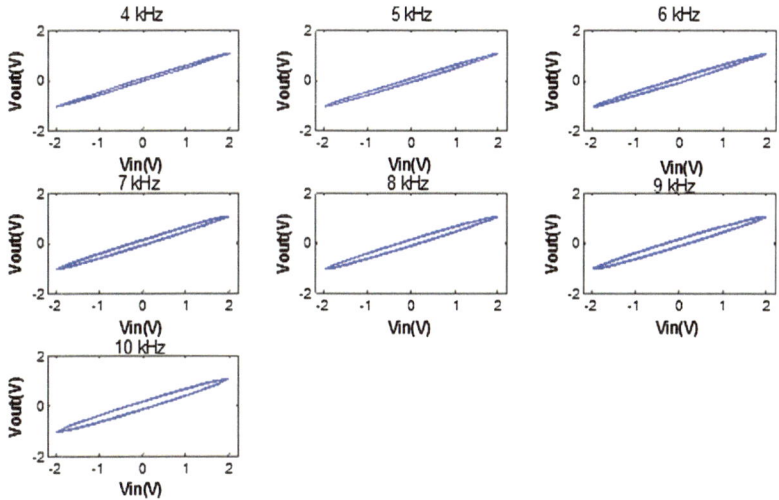

Fig. 2.18 Lissajous curves on measured input V_{in} and output V_{in} voltages in the frequency range between 4 and 10 kHz for the device IPMC$_{AbsT-5h}$

The IPMC$_{AbsT-10h}$ and IPMC$_{AbsT-20h}$ show the same trend in the Lissajous curves, and therefore the conclusion about IPMC$_{AbsT-5h}$ linearity will be extended to them.

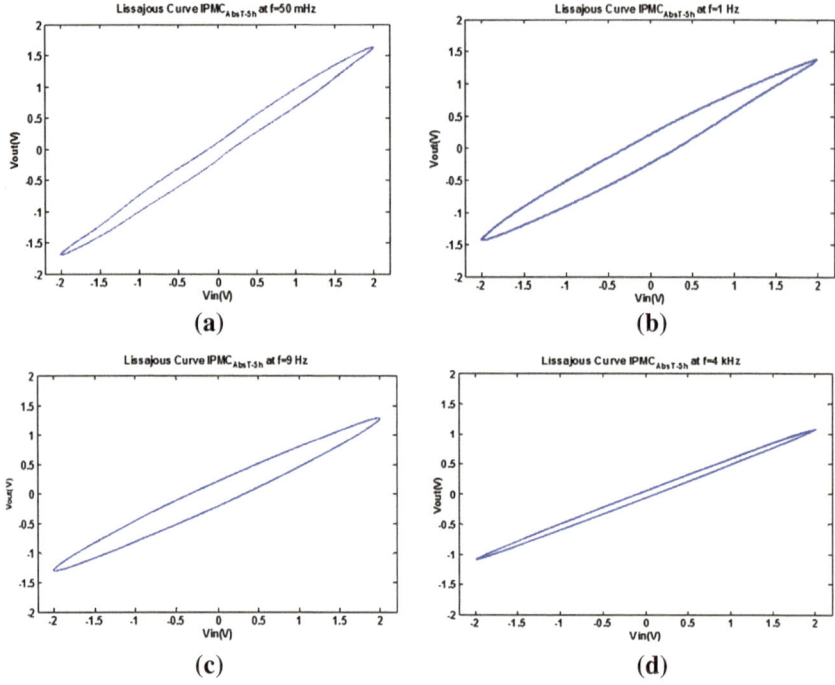

Fig. 2.19 Shape analysis of Lissajous curves on measured input V_{in} and output V_{in} voltages for the device IPMC$_{AbsT-5h}$: **a** non-elliptic shape at $f = 50$ mHz, **b** transition between non-elliptic to elliptic shape at $f = 1$ Hz, **c** elliptic shape at $f = 9$ Hz, **d** elliptic shape at $f = 4$ kHz

At low frequencies, a capacitance is considered an open circuit and the nonlinearity dominates the global behavior. As the frequency increases, the linear capacitive effect of IPMC becomes significant and dominant with respect to the nonlinear component.

Concluding, in this stage, IPMC FOE will be approximated as linear in a frequency range from 1 Hz to 10 kHz.

2.4.2.1 IPMC Frequency Analysis

The modulus and phase diagrams, see Figs. 2.20, 2.21, and 2.22 of the ratio between V_{out} and V_{in} signals at each frequency, have been obtained using the experimental data in the frequencies range between 10 mHz and 10 kHz for all the devices IPMC$_{AbsT-5h}$, IPMC$_{AbsT-10h}$, and IPMC$_{AbsT-20h}$.

Given the conclusion on linearity assessed by the Lissajous curves, such curves can be considered as the system's frequency response in terms of Bode diagrams of the transfer function in the linearity range from 1 Hz to 10 kHz. In such a range, it was observed that both IPMC devices show a fractional-order behavior in a limited span of frequencies where the modulus of the Bode diagrams presents a slope equal to $\alpha \times$

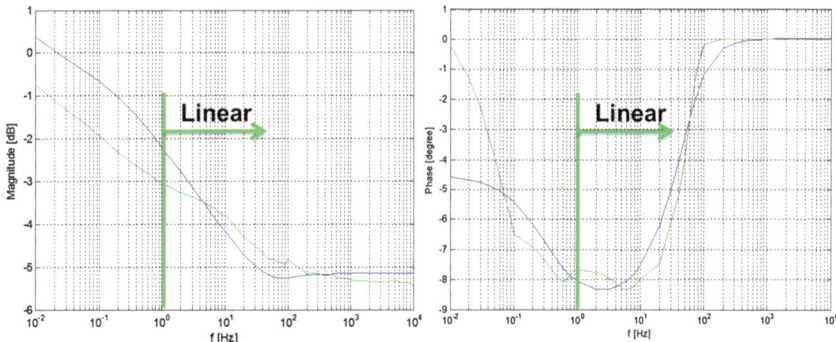

Fig. 2.20 Modulus (**a**) and phase (**b**) diagrams of the ratio $\frac{V_{out}}{V_{in}}$ in the complete experimental frequencies range: 10 mHz to 10 kHz for the IPMC device at 5 h absorption time

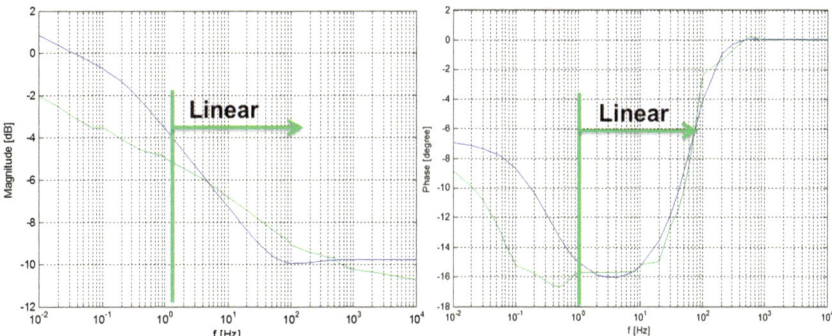

Fig. 2.21 Modulus (**a**) and phase (**b**) diagrams of the ratio $\frac{V_{out}}{V_{in}}$ in the complete experimental frequencies range: 10 mHz to 10 kHz for the IPMC device at 10 h absorption time

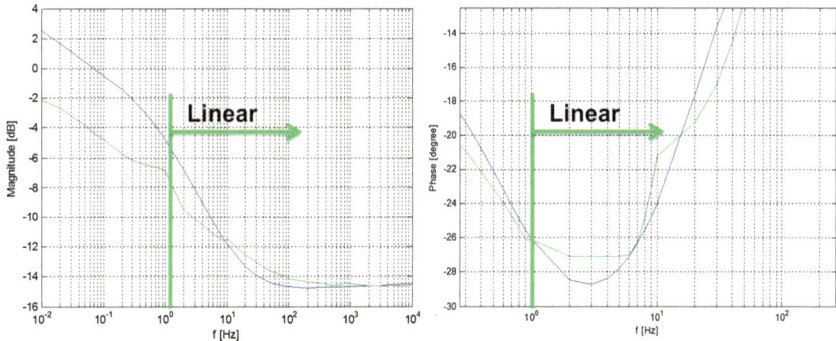

Fig. 2.22 Modulus (**a**) and phase (**b**) diagrams of the ratio $\frac{V_{out}}{V_{in}}$ in the complete experimental frequencies range: 10 mHz to 10 kHz for the IPMC device at 20 h absorption time

20 db/decade, and the phase presents a lag equal to $\alpha \times 90°$, where α is identifiable as the fractional-order exponent.

With respect to the three IPMC devices, the following comparisons can be made:

- IPMC$_{AbsT-5h}$: IPMC device with 5 h of absorption time
  The IPMC with 5 h of absorption time has shown an average slope in the modulus diagram of 1 dB per decade in the frequency range between 1 and 10 Hz, determining $\alpha = 0.05$. The phase diagram showed an average phase of $-4.5°$ in the same frequency range being coherent with the fractional exponent related to the modulus $-\alpha \times 90° = -4.5°$.
- IPMC$_{AbsT-10h}$: IPMC device with 10 h of absorption time
  The IPMC with 10 h of absorption time has shown an average slope in the modulus diagram of 1.6 dB per decade in the frequency range between 1 and 10 Hz, determining $\alpha = 0.075$. The phase diagram showed an average phase of $-6.7°$ in the same frequency range being coherent with the fractional exponent related to the modulus $-\alpha \times 90° = -6.7°$.
- IPMC$_{AbsT-10h}$: IPMC device with 20 h of absorption time
  The IPMC with 20 h of absorption time showed an average slope in the modulus diagram of 6 dB per decade in the frequency range between 1 and 10 Hz, determining $\alpha = 0.3$. The phase diagram showed an average phase of $-27°$ in the frequency range between 1 and 10 Hz showing, in that range, coherence with the fractional exponent related to the modulus $-\alpha \times 90° = 27°$.

Finally, Fig. 2.23 shows the trend of the fractional-order α while varying the absorption time.

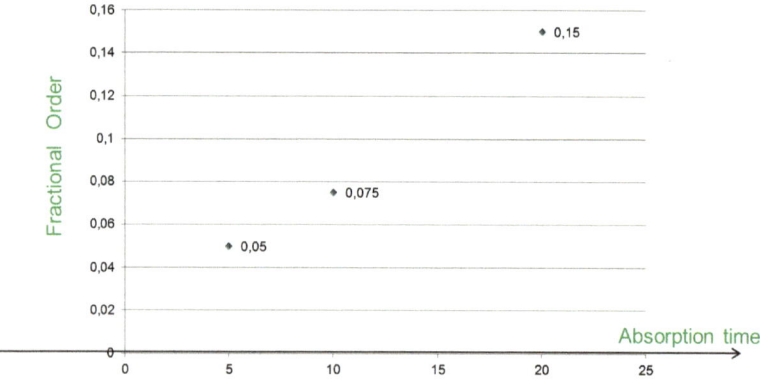

Fig. 2.23 Fractional-order versus absorption time

2.4.3 Carbon Black-Based FOE

The focus of this paragraph is on the possibility to exploit nanostructured material, in particular carbon black (*CB*)-based composite, to realize non-integer order devices. Such elements can be modeled as FOE whose experimentally acquired frequency response fit the following model:

$$Z(s) = \frac{K}{(1 + \tau s)^\alpha}. \tag{2.2}$$

It is known in the literature that by adding a conductive filler to an insulating matrix, generally a polymeric matrix, both the electrical and mechanical properties of the composite can change. Such phenomenon has been studied for realizing polymer conducting composites, with application, such as thermistors, deformation sensors [23], pressure sensors [20], and gas sensors [50].

Here, the possibility to use carbon black to realize nanocomposites with dielectrics properties is investigated.

2.4.3.1 Materials

Sylgard$_{184}$, the host matrix of interest in this research, is a thermally curable silicone rubber.

Its formulation consists of two parts: vinyl-terminated polydimethylsiloxane (PDMS) chains (part A) and a mixture of methylhydrosiloxane copolymer chains with a Pt catalyst and an inhibitor (part B).

Parts A and B are viscous liquids, 5000 cSt and 110 cSt at 25 °C, respectively, usually mixed together in a 10:1 ratio.

Addition of a catalytic amount of the SiH of methylhydrosiloxane (part B) to the double bond of the vinyl-terminated PDMS species (part A) leads to a cross-linked network through the formation of Si–C covalent bonds.

The described, nanomaterial CB-based polymer, is in fact the dielectric of the CB–FOE under investigation. The resulting structure of the CB–FOE is given in Fig. 2.24.

The obtained dielectrics have been used to realize cylindrical capacitors, whose geometry is shown in Fig. 2.25. More specifically, capacitances considered in the following had copper-based shell with height $h = 8$ cm, internal diameter $a = 0.6$ cm, and external diameter $b = 1.2$ cm.

2.4.4 FOE Under Test and Experimental Setup

The samples have been prepared by mixing the PDMS and the cross-linking agent in a ratio of 1:10 in a Teflon crucible.

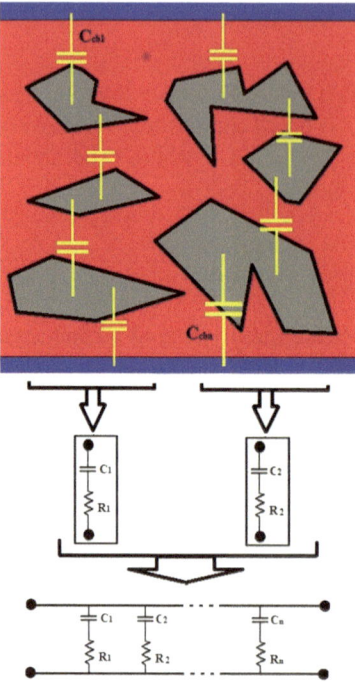

Fig. 2.24 Carbon black-based fractional-order element (CB-FOE) structure

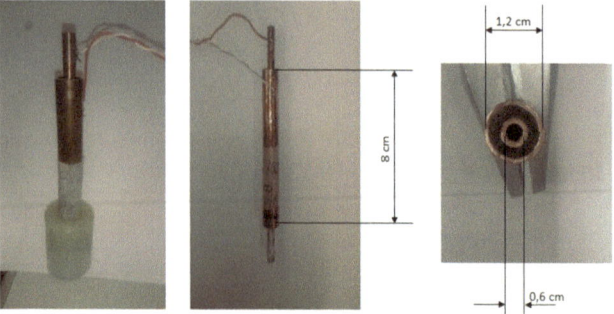

Fig. 2.25 Sample of the CB-FOE under investigation

The mixture was mixed for 10 min and at this point has been added the required amount of carbon black to achieve the desired concentration.

It is continued to stir for 10 min to enhance the dispersion of the carbon black, and then the viscous mixture was poured into the device and it is made reticular at 25 °C for 2 days or at 150 °C for 28 min.

Table 2.1 List of the realized CB-FOE

Carbon black %	Curing temperature	Characteristic (pF)
1%	Room	≈25
2%	Room	≈30
8%	100 °C	≈76.6
8%	125 °C	≈207
8%	150 °C	≈150
5% xiline solvent	Room	≈137
5% CHCL3 solvent	Room	≈50

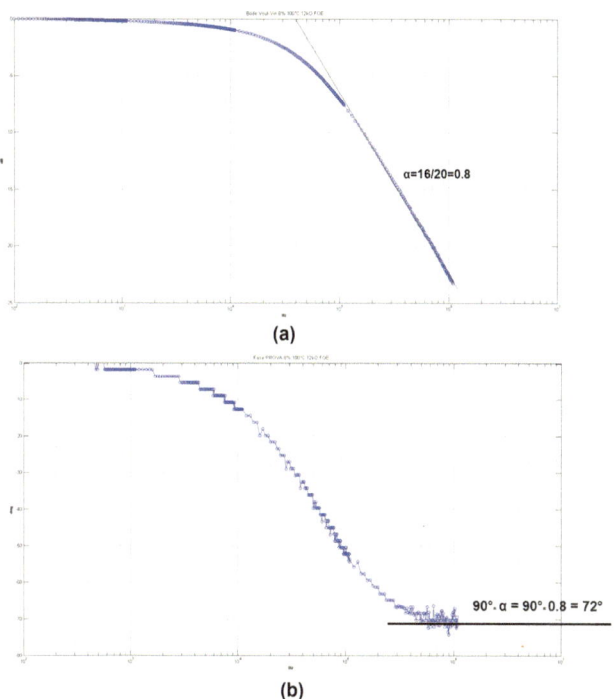

Fig. 2.26 Frequency-domain identification, device 8%@100 °C

The carbon black percentage, the curing temperature, and the solvent type have been changed according to Table 2.1. The capacitors, during the acquisition phase, were connected in series configuration with a resistor of 12 kΩ.

A sinusoidal input voltage V_{in} was produced using a waveform generator Agilent 33220A. The sinusoidal signal was applied through an operational amplifier, ST TL082CP, in buffer configuration.

The investigated frequency range was from 10^2 to 10^7 Hz; moreover, 100 points per decade were considered, see Figs. 2.26, 2.27, and 2.28.

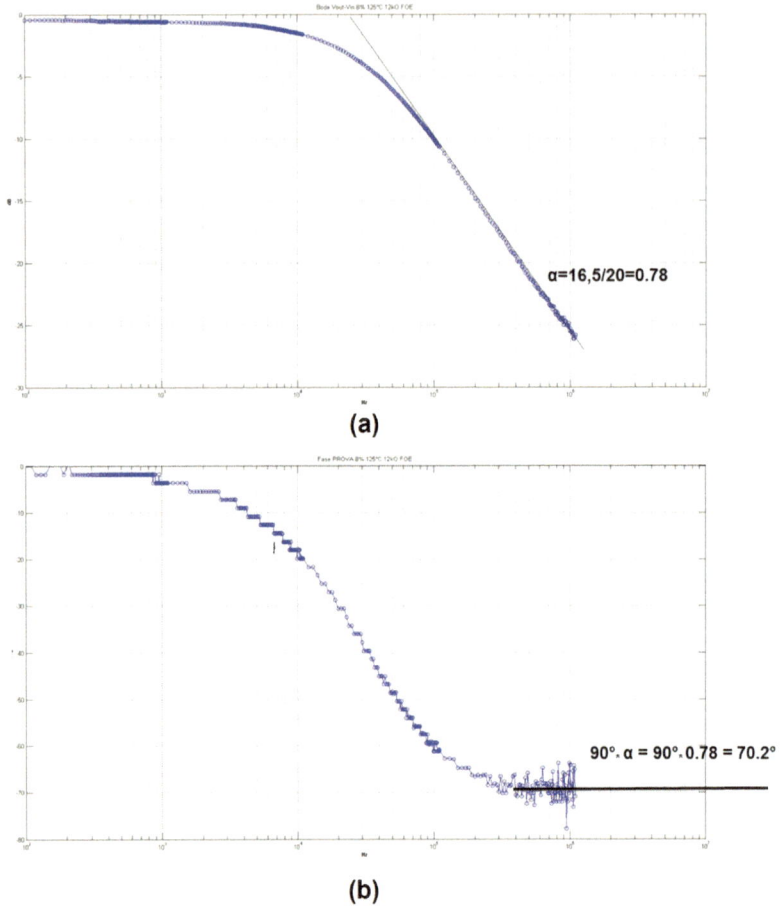

Fig. 2.27 Frequency-domain identification, device 8%@125 °C

The obtained result allows to foresee a dependence of the order α with the curing temperature.

In fact, in Table 2.2 it is possible to note that α decreases from 0.8 to 0.7 via 0.78 when the curing temperature raises from 100 to 150 °C.

2.5 Other Solid-State Devices

In addition to the fractional-order elements mentioned above, a few other solid states devices have been reported in literature. One such system is based on the *Fractal structure*. In 1997, Haba et al. developed a fractal structure with three levels of iterations which gives a prominent CP zone [24]. Later, they studied this fractal

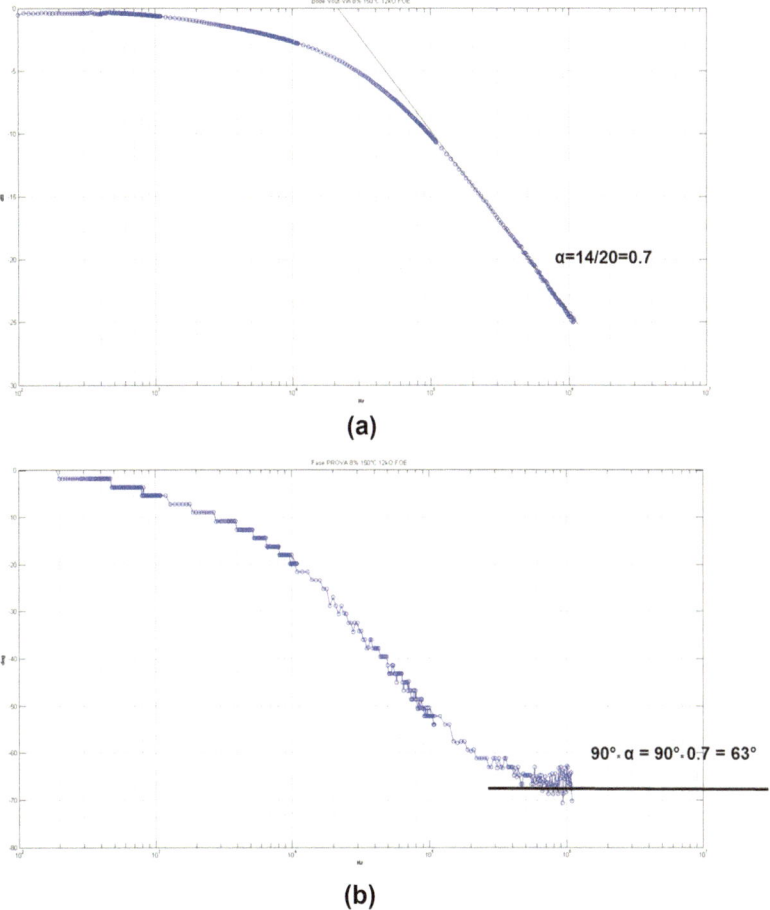

Fig. 2.28 Frequency-domain identification, device 8%@150 °C

structure for varying fractal resistances and capacitances and showed how CP zone can be varied using those resistance and capacitance [25]. It shows CP in 10^5 Hz to 10^{10} Hz zone with CP angle = 36° (i.e., $\alpha = 0.4$) and fabricated by photolithography using Si substrate. The reported CPZs are usually three to five decades long. These devices are lightweight and dry fractors with a phase ripple of $<\pm 2.7°$.

Another solid-state FOE worth to be mentioned is based on a graphene–polymer composite. Here, fractal or porous structures were introduced to bring the effect of dispersive RC blocks, as pursued in different multicomponent techniques. In this work, dielectric of conventional parallel-plate capacitor is replaced by a polymer composite, percolated with reduced graphene oxide [21]. The properly dispersed graphene sheets in the polymer gives a number of RC blocks of different relaxation times, which ultimately give FO nature. This FOE shows different α values (0.33–

Table 2.2 Determination of the α value using the asymptotic phase calculus

Curing temperature (°)	Asymptotic value (deg)	α	Magnitude slope (dB/dec)
100	$\simeq$-72	$\simeq$0.8	$\simeq$16
125	$\simeq$-70.2	$\simeq$0.78	$\simeq$16.5
150	$\simeq$-63	$\simeq$0.7	$\simeq$14

0.73) for different graphene loads (12%–2% respectively). This device is small in size, compatible to microfabrication, but could be realized for very small CPE zone (400 kHz–2 MHz only). The fabrication procedure is complicated and costly.

2.6 Solution-Based Systems

There has been significant research in electrochemistry verifying the validity of the CPE model describing embedded dynamics. The performance of these materials, depending on the processes of the electrolytes, has been described, and the physical origins of the model parameter have been established in many cases. However, the influence of the electrodes was not a subject of study and only recently this problem became relevant in the viewpoint of fractional calculus.

In this section, fabrication details of porous polymer-based electrolytic fractor and its potential for use in realizing fractional-order systems will be discussed. Our efforts in solution-based FOE were preceded by Jesus and Machado [26].

Fractal electrode-based electrolytic FOE: In 2008, Jesus and Machado [26] used the concepts of using both the fractal structure and electrolyte to realize FOE. Here, electrodes are Cu-clad PCB with a typical fractal geometry (carpet of Sierpinski) and electrolyte is NaCl solution. A sand stand within the NaCl solution is used as the fractal material. It is shown how α increases with the increase of effective area of fractal structure as well as solution molarity. Here, α values from 0.2 to 0.6 have been realized. The main limitation is the bulky structure and potential for spillable; thus limiting application in various FO systems.

Porous polymer-based electrolytic fractor: This prototype device was first reported by Biswas et al. [8]. It is made of porous PMMA-coated Cu or Pt electrodes, dipped in an ionic solution. Later, such fractors are packaged by replacing the ionic solution by ionic gel (made by agar-agar powder) [31]. Different α values between 0.1 and 0.85 have been achieved. The realized FOE covers the frequency zone from 100 Hz to 1 MHz. Initially, the constant phase zone (CPZ), i.e., the bandwidth of the FOE was maximum two decades long [30]. Later by introducing the CNT in the coating polymer (in this case coating polymer is BPADA–mPD) the bandwidth could be extended to five decades (20 Hz–2 MHz) [1]. The packaged FOE is smaller than the fractal FOE, cheaper, and simpler to fabricate. It is successfully used in developing many FO systems [8, 31, 45–47]. This FOE has been studied for different system parameters, i.e., dipping length, coating thickness, the solution pH, and conductivity. It is shown that CP zone can be moved to high-frequency zone by increasing solution pH or conductivity, and CPZ can be widened by decreasing coating thickness. Though a definite equation between α and the system parameters could not be established yet, the study elaborates how we can design the system parameter to get a particular value of α and are detailed in the next section.

2.6.1 Fabrication Details of the Solution-Based FOE

It has been reported in the literature that when a porous electrode is dipped in an electrolyte then diffusion takes place through the electrode surface. If the surface is smooth then ordinary diffusion takes place; that is, the mean free path of the diffusing particles follows the ordinary linear law of diffusion [6],

$$r^2 \propto t, \tag{2.3}$$

where r is the mean free path displacement and t is the time constant. But for a porous surface the above-stated ordinary diffusion equation does not hold accurately. For this anomalous diffusion to take place, the mean square displacement of the diffusing particles shows a PL distribution on time as expressed by

$$r^2 \propto t^{\alpha}, \tag{2.4}$$

where α is fractional in nature. Considering the Fick's law of diffusion, continuity equation, and modified laws for anomalous diffusion, the impedance of diffusion process can be modeled by "Transmission Line" which has the similarity with the infinite RC ladder network mentioned by the researchers in the field of circuit theory [27, 33, 39] to realize FOE.

To get the porous surface, a thin coating (5–50 μm) was provided by dipping the probe in PMMA solution and the probes exhibit constant phase behavior with an angle much less than 90° when dipped in ionic solution. This can be explained as the relaxation time at the metal–PMMA electrolyte interface gives a fractional exponent α much less than the value 1. The probe shown in Fig. 2.29 is made of about 6-mm-wide strip cutout from about 1.6-mm-thick plate generally used as double-sided printed circuit boards (PCB). The copper claddings on the two external faces serve as the two electrodes of the probe. The construction of PCB ensures sufficient rigidity to the probe. The fabrication of PMMA-coated probes is reported in [9]. For 1% PMMA solution, 0.44 g of PMMA flex is fully dissolved in 30 mL of chloroform to prepare the coating solution. The copper-coated epoxy strip is then dipped into this solution. Thus, a layer of PMMA solution is coated over the copper surface.

The insulation process of the metal electrodes of the probe by the above-mentioned procedure forms a thin porous film of PMMA on the metal electrodes. This may be

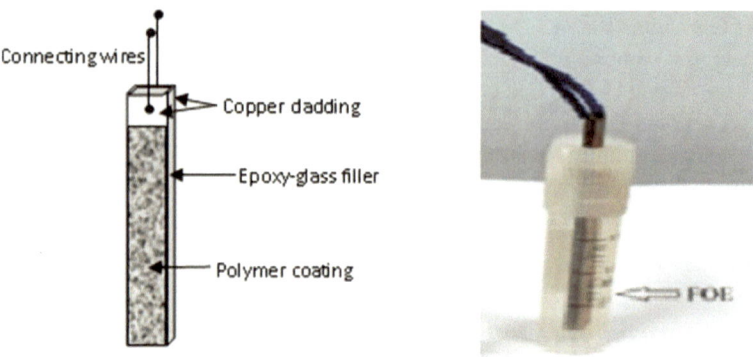

Fig. 2.29 Schematic of the polymer-coated probe and the solution-based FOE

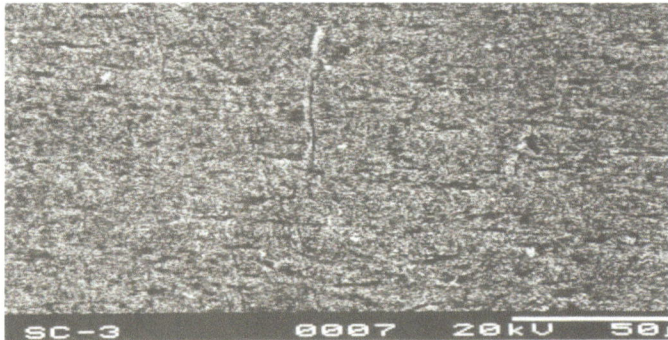

Fig. 2.30 The SEM picture of the surface of copper electrode of the probe without coating

due to the fact that when the solvent chloroform evaporates, nanopores are formed on the surface. This can be seen in the scanning electron microscope (SEM) pictures of the surface of the copper electrodes taken before coating and after the coating by PMMA. Figure 2.30 shows a SEM picture of the bare surface of the electrodes at a magnification of 500 and Fig. 2.31 shows the same for a PMMA-coated surface with coating thickness of 5 µm. The regular porous nature of the surface is similar to the "Transmission Line" model as described by Bisquert et al. [7] and results "CPA" behavior of the process. The CPA behavior for such probes are shown in Figs. 2.33 and 2.34.

In place of PMMA other polymer can also be used if they are soluble in a solvent which evaporates during drying, leaving the porous polymer surface on the electrodes. One such polymer is BPADA–mPD. It is a composition of 4,4'-(4,4'-isopropylidene di phenoxy) bis (phthalic anhydride) (BPADA) and m-phenylene diamine (mPD). The FOE realized using BPADA–mPD is elaborated later.

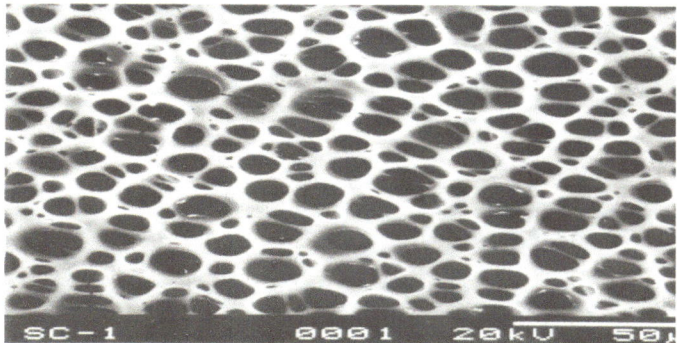

Fig. 2.31 The SEM picture of the surface of PMMA coating of 5 μm (2.5%) thickness at a magnification of 500

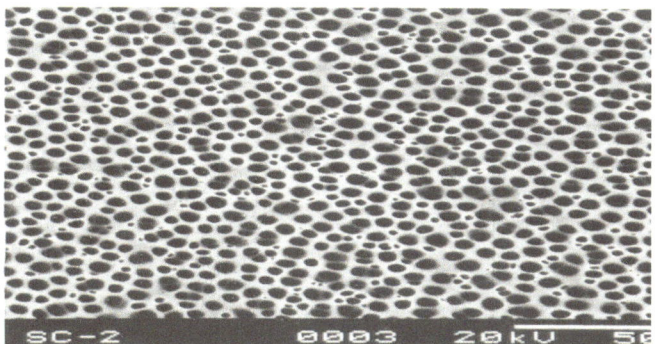

Fig. 2.32 The SEM picture of the surface of PMMA coating of about 12 μm (5%) thickness at a magnification of 500

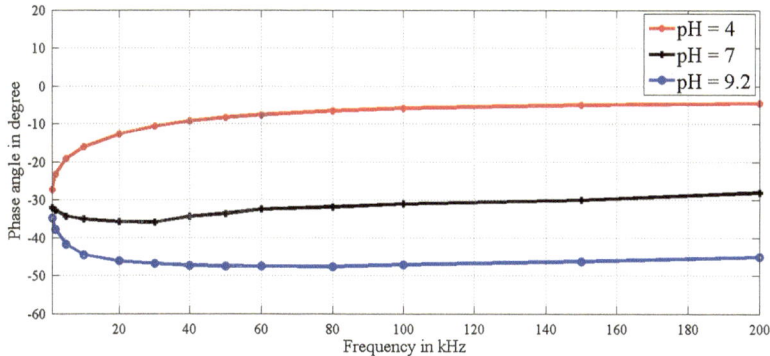

Fig. 2.33 Constant phase angle of the FOE realized by dipping the PMMA-coated probe in three buffer pH solutions

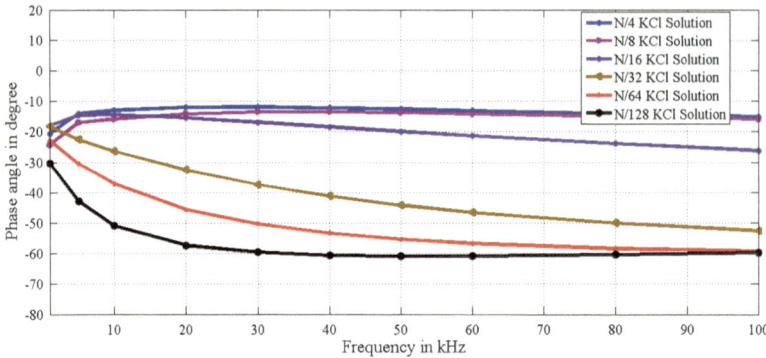

Fig. 2.34 Constant phase angle of the FOE realized by dipping the PMMA-coated probe in six different aqueous solutions of KCl

2.6.2 The Parameter Dependence of the FOE

The characteristics of FOE are mainly dependent upon the fractional exponent α in (1.1), that means on the constant phase angle θ which is being measured by the LCR meter. The constant phase angle depends on a number of parameters. This has been investigated experimentally and from the experimental observation we can say

$$\theta_{FOE} = f(t, A, \sigma), \tag{2.5}$$

where t is the thickness of the insulation on the electrode, A is the area of contact of the electrodes with the polarizing medium, and σ is the ionic concentration of the polarizing medium.

2.6.2.1   Dependence of Phase Angle on the Ionic Concentration, "σ"

The exponent α of (1.1) varies with the ionic concentration of the solution in which the polymer-coated probe is dipped. This is illustrated by the graphs of Fig. 2.33. It can be seen from the figures that when the same probe is dipped in three different buffer solutions of pH (pH 4, pH 7, and pH 9.2), they result in different phase angles and hence different values of the exponent, α. Similar observation also can be found in Fig. 2.34, wherein different concentrations of KCl solution result in different phase angles.

2.6.2.2   Dependence of Phase Angle on "t"

The fractional exponent α is very much dependent upon the coating thickness of the porous polymer on the surface of the metal electrode. It has been observed

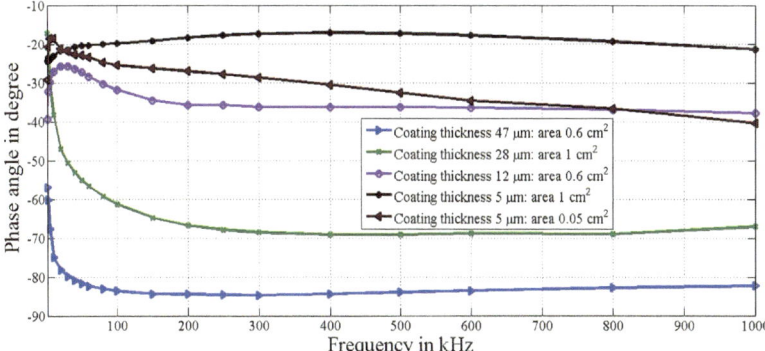

Fig. 2.35 Behavior of constant phase angle characteristics of probes constructed with varying coating thickness, length, and width when immersed 1 cm in ionic medium (condutivity = 250μS/cm); for coating thickness = 47 μm, area in polarizable medium is 0.6 cm^2; for coating thickness = 28 μm, area in polarizable medium is 1 cm^2; for coating thickness = 12 μm, area in polarizable medium is 0.6 cm^2; for coating thickness = 5 μm, area in polarizable medium is 1 cm^2; for coating thickness = 5 μm, area in polarizable medium is 0.05 cm^2

that for a thick coating it acts as a capacitor with the phase angle near about 90° and as the coating thickness decreases the fractional exponent also decreases and gives lower phase angle in the impedance measurement of the probe by LCR meter. Figure 2.35 shows the dependence of the phase angle on the thickness of the coating. For coating thickness 47 and 12 μm, the area in contact with the polarizing medium is same but different coating thicknesses results in a wide difference in the measured constant phase angle. The same observation is also true for coating thickness of 28 and 5 μm. This is because, with the increased number and size of pores (for lesser coating thickness); the electric current density takes less time to follow the change of the electric field in the dielectric region. As a result, the phase angle of the impedance is less than the case when the size and number of the pores are less. This phenomenon can be further illustrated by the "*SEM*" picture taken for two different coating thicknesses and shown in Figs. 2.31 and 2.32. So, from the above observations, it can be said that coating thickness becomes a design parameter for realizing different FOEs.

2.6.2.3 Dependence of Phase Angle on "*A*"

The dependence of constant phase angle, θ, on the area of contact with the polarizing medium can be observed in Fig. 2.35. From the graphs of Fig. 2.35, it is apparent that when the coating thickness is 5 μm the CPA is nearly 30° for 1 cm^2 contact area and CPA is nearly 20° for contact area of 0.05 cm^2; thus the area of contact can be varied to get different FOEs.

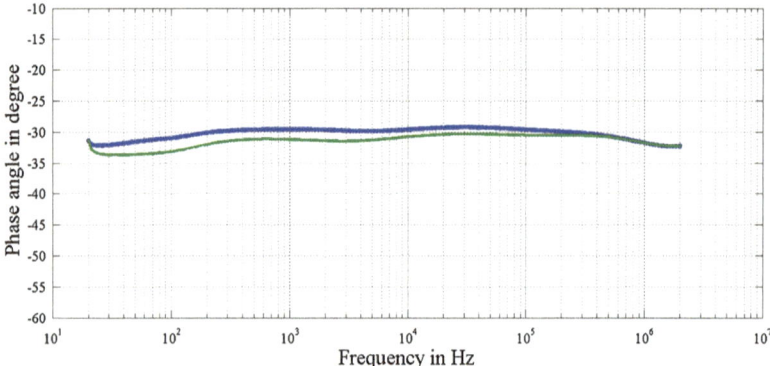

Fig. 2.36 Constant phase angle of the FOE realized by dipping the probe coated with BPADA–mPD + 1%CNT in pH 9.2 buffer solution

2.6.3 CNT–Polymer Composite-Based Wideband FOE

This wideband FOE was reported in [1]. It has similar structure like that of solution-based system discussed above [8], the difference is that the electrodes are now coated with carbon nanotube (CNT)–polymer composite instead of porous PMMA. The coating polymer is BPADA–mPD. It is a composition of 4,4'-(4,4'-isopropylidene di phenoxy) bis (phthalic anhydride) (BPADA) and m-phenylene diamine (mPD). Carbon nanotube (CNT) has been introduced to coating polymer, to achieve dispersive resistive–capacitive blocks, that is, the physics behind the realization technique of ladder-realized fractors. That means, now, CNT loading is also a design parameter which enhances the design flexibility [1].

In PMMA-based FOE, the effect of porosity does not become prominent until the coating thickness is low. So, the bandwidth is small in [8]. This problem is overcome by adding 1.5% CNT loading in the polymer. This CNT-based FOE shows five decades long bandwidth ranging from 20 Hz to 2 MHz with $\alpha = 0.35$. The main advantages are as follows: long bandwidth, low-phase ripple ($\pm 2°$ only), and high yield rate. The main limitation is that any variation in α is yet not explored in this work. Phase plot of one such FOE is shown in Fig. 2.36.

2.6.4 Fractance-Based Sensor

It will be worth to mention here that the FOE can be used for sensing purposes by correlating the phase angle with the concentration of components in the electrolyte.

In Sect. 2.6.2, it is mentioned that the phase angle of a FOE changes with the change of ionic concentration of the polarizing medium. This change of phase angle behavior leads to the concept of a new type of sensor, where the phase angle changes with the process variables. Characteristics of a fractance sensor differ from a conventional "Impedance Sensor" in a way that the phase angle introduced by the sensor is measured (as in the present case); as a result the measuring circuit becomes simpler. A signal conditioning circuit, based on phase detection, can be developed which gives linear output with the phase angle change. Discussion on the sensing phenomenon based on the fractance properties is an active area of research and may help in our understanding of how to model time-varying order parameters.

2.7 Lesson Learned

This chapter introduced a few approaches to developing FO devices. The most important point is that we can use the ubiquity of the "universal dielectric response" to create such devices using different materials and methods, solid state and liquid based, all well modeled by the fractional calculus. In the next chapter, we see a sampling of the applications of such devices in actual physical demonstrations. In the following chapter, we make connections with living organic matter, looking at the kind of FO dynamics nature creates.

References

1. A. Adhikary, M. Khanra, S. Sen, K. Biswas, Realization of a carbon nanotube based electrochemical fractor, in *IEEE International Symposium on Circuits and Systems (ISCAS'15)*, Lisbon, Portugal, May (2015), pp. 2329–2332
2. A. Adhikary, P. Sen, S. Sen, K. Biswas, Design and performance study of dynamic fractors in any of the four quadrants. Circuits Syst. Signal Process. **35**(6), 1909–1932 (2015)
3. A. Adhikary, G. Kumar, S. Bannerje, S. Sen, K. Biswas, Modelling and performance improvement of phase-angle-based conductivity sensor, in *IEEE First International Conference on Control, Measurement and Instrumentation (CMI)*, Kolkata, India, Jan (2016), pp. 403–407
4. A. Adhikary, S. Sen, K. Biswas, Practical realization of tunable fractional order parallel resonator and fractional-order filters. IEEE Trans. Circuits Syst. I **63**(8) (2016)
5. E. Barsoukov, J.R. Macdonald, *Impedance Spectroscopy: Theory, Experiment and Applications*, 2nd edn. (Wiley, Hoboken, New Jersey, 2005)
6. J. Bisquert, A. Compte, Theory of electrochemical impedance of anomalous diffusion. J. Electroanal. Chem. **499**, 112–120 (2001)
7. J. Bisquert, G. Garcia-Belmonte, F. Fabregat-Santiago, N.S. Ferriols, P. Bogdanoff, E.C. Pereira, Doubling exponent models for the analysis of porous film electrodes by impedance: relaxation of shape ti shape o_2 nanoporous in aqueous solution. J. Phys. Chem. **104**, 2287–2298 (2000)
8. K. Biswas, S. Sen, P.K. Dutta, Realization of a constant phase element and its performance in a differentiator circuit. IEEE Trans. Circuits Syst. II **53**(9), 802–806 (2006)
9. K. Biswas, Studies on design, development and performance analysis of capacitive type sensors. Ph.D. Thesis, Indian Institute of Technology Kharagpur, India, Department of Electrical Engineering—India (2007)
10. H. Bode, *Network Analysis and Feedback Amplifier Design* (Van Nostrand, New York, 1945)
11. C. Bonomo, L. Fortuna, P. Giannone, S. Graziani, S. Strazzeri, A nonlinear model for ionic polymer metal composites as actuators. Smart Mater. Struct. **16**, 1–12 (2007)

12. R. Caponetto, D. Porto, Analog implementation of non integer order integrator via field pro-
    grammable analog array, in *FDA'06: Proceedings of the 2nd IFAC Workshop on Fractional
    Differentiation and Its Applications*, Porto, Portugal (2006), pp. 170–173
13. R. Caponetto, S. Graziani, F.L. Pappalardo, F. Sapuppo, Experimental characterization of ionic
    polymer metal composite as a novel fractional-order element. Adv. Math. Phys. **2013**, 1–10
    (2013)
14. R. Caponetto, D.G. Dongola, L. Fortuna, A. Gallo, New results on the synthesis of FO-PID
    controllers. Commun. Nonlinear Sci. Numer. Simul. **15**, 997–1007 (2010)
15. R. Caponetto, G. Dongola, L. Fortuna, S. Graziani, S. Strazzeri, A fractional model for IPMC
    actuators, in *IEEE International Instrumentation and Measurement Technology Conference*,
    Canada (2008)
16. Y.-Q. Chen, K.L. Moore, Discretization schemes for fractional-order differentiators and inte-
    grators. IEEE Trans. Circuits Sys.–I: Fund. Theory Appl. **49**(3), 363–367 (2002)
17. Y.-Q. Chen, B.M. Vinagre, I. Podlubny, Continued fraction expansion approaches to discretiz-
    ing fractional-order derivatives—an expository review. Nonlinear Dyn. **38**, 155–170 (2004)
18. W.S. Chu, K.T. Lee, Review of biomimetic underwater robots using smart actuators. Int. J.
    Precis. Eng. Manuf. **7**, 1281–1292 (2012)
19. S. Cole, R. Cole, Dispersion and absorption in dielectrics. J. Chem. Phys. **9**, 341–351 (1941)
20. M. Ding, L. Wang, P. Wang, Changes in electrical resistance of carbon-black filled silicone
    rubber composite during compression. J. Polym. Sci. Part B: Polym. Phys. **45**(19), 2700–2706
    (2007)
21. A.M. Elshurafa, M.N. Almadhoun, K.M. Salama, N.H. Alshareel, Microscopic electrostatic
    fractional capacitors using reduced graphene oxide percolated polymer composites. Appl. Phys.
    Lett. **102**, (232901(4 pp)) (2013)
22. M. Furlani, M.C. Stappen, B.E. Mellander, G.A. Niklasson, Concentration dependence of ionic
    relaxation in lithium doped polymer electrolytes. J. Non-Cryst. Solids **356**, 710–714 (2010)
23. P. Giannone, S. Graziani, E. Umana, Investigation of carbon black loaded natural rubber
    piezoresistivity, in *Proceedings of IEEE I2MTC* (2015), pp. 1477–1481
24. T.C. Haba, G. Ablart, T. Camps, The frequency response of a fractal photolithographic structure.
    IEEE Trans. Dielectr. Electr. Insul. **4**(3), 321–326 (1997)
25. T.C. Haba, G. Ablart, T. Camps, F. Olivie, Influence of the electrical parameters on the input
    impedance of a fractal structure realised on silicon. Chaos, Solitons Fractals **24**, 479–490 (2005)
26. T.S. Jesus, J.A.T. Machado, Development of fractional-order capacitors based on electrolyte
    processes. Nonlinear Dyn. **56**(1), 45–55 (2009)
27. R.M. Lerner, The design of a constant-angle or power-law magnitude impedance. IEEE Trans.
    Circuit Theory 98–107 (1963)
28. C. Longfei, C. Hualing, Z. Zicai, A structure model for ionic polymer-metal composite (IPMC),
    in *Proceedings of SPIE 8340, Electroactive Polymer Actuators and Devices (EAPAD)* (2012)
29. D. Mondal, K. Biswas, Performance study of fractional order integrator using single-component
    fractional-order element. IET Circuits, Devices, Syst. **5**(4), 334–342 (2011)
30. D. Mondal, Fabrication and performance studies of PMMA—coated fractional-order elements,
    Master's thesis, Indian Institute of Technology Kharagpur, India, Department of Electrical
    Engineering, Nov (2012)
31. D. Mondal, K. Biswas, Packaging of single component fractional-order element. IEEE Trans.
    Device Mater. Rel. **13**(1), 73–80 (2013)
32. C.A. Monje, Y.-Q. Chen, B.M. Vinagre, D. Xue, V. Feliu, *Fractional-Order Systems and Con-
    trols: Fundamentals and Applications*. A Monograph in the Advances in Industrial Control
    Series (Springer, Berlin, 2010)
33. K.B. Oldham, Semintegral electroanalysis: analog implementation. Anal. Chem. **45**, 39–47
    (1973)
34. A. Oustaloup, P. Lanusse, P. Melchior, X. Moreau, J. Sabatier, J.L. Thomas, A survey of the
    CRONE approach, in *Conference Proceedings 1st IFAC Workshop on Fractional Differentiation
    and its Applications FDA'04* (Bordeaux, FR, 2004)

35. P. Pasierb, S. Komornicki, R. Gajerski, S. Koziński, P. Tomczyk, M. Rekas, Electrochemical gas sensor materials studied by impedance spectroscopy, Part I: Nasicon as a solid electrolyte. J. Electroceram. **8**, 49–55 (2002)
36. I. Petráš, B.M. Vinagre, Practical application of digital fractional-order controller to temperature control. Acta Montanist. Slovaca **7**, 131–137 (2002)
37. I. Podlubny, I. Petráš, B.M. Vinagre, P. O'Leary, L. Dorčák, Analog realizations of fractional-order controllers. Nonlinear Dyn. **29**, 281–296 (2002)
38. D. Pugal, K.A. Jung, A. Aabloo, K.J. Kim, Ionioc polymer-metal composite mechanoelectrical tranduction: review and perspectives. Polym. Int. **59**, 279–289 (2009)
39. S.C.D. Roy, On the realization of a constant-argument immitance or fractional operator. IEEE Trans. Circuit Theory **CT-14**, 264–274 (1967)
40. H. Samavati, Fractal capacitors. IEEE J. Solid-State Circuits **33**(10), 2035–2041 (1998)
41. M. Shahinpoor, J. Kim, Ionic polymer-metal composites: I. Fundamentals. Smart Mater Struct. **10**, 819–33 (2001)
42. M. Shahinpoor, Y. Bar-Cohen, Y. Simpson, J. Smith, Ionic polymer-metal composites (IPMC) as biomimetic sensors, actuators, and artificial muscle—a review. Smart Mater. Struct. **7**, R15–30 (1998)
43. M. Shainpoor, K.J. Kim, Ionic polymer-metal composites: IV. Industrial and medical applications. Smart Mater. Struct. **14**, 197–214 (2005)
44. D. Sierociuk, I. Podlubny, I. Petráš, Experimental evidence of variable-order behavior of ladders and nested ladders. IEEE Trans. Control Syst. Tech. **21**(2), 459–466 (2013)
45. M.C. Tripathy, D. Mondal, K. Biswas, S. Sen, Design and performance study of phase-locked loop (shape PLLS) using fractional-order loop filter. Int. J. Circuit Theory Appl. **43**(6), 776–792 (2015)
46. M. Tripathy, K. Biswas, S. Sen, A design example of a fractional-order Kerwin-Huelsman-Newcomb biquad filter with two fractional capacitors of different order. Circuits, Syst. Sig. Process. **32**(4), 1523–1536 (2013)
47. M.C. Tripathy, D. Mondal, K. Biswas, S. Sen, Experimental studies on realization of fractional inductors and fractional-order bandpass filters. Int. J. Circuit Theory Appl. **43**(9), 1183–96 (2015)
48. G. Tsirimokou, C. Psychalinos, A. Elwakil, Digitally programmed fractional-order Chebyshev filters realizations using current-mirrors, in *2015 International Symposium on Circuits and Systems (ISCAS'15)*, 24–27 May 2015, Lisbon, pp. 2337–2340
49. B.M. Vinagre, Y.-Q. Chen, I. Petráš, Two direct Tustin discretization methods for fractional-order diferentiator/integrator. J. Franklin Inst. **340**, 349–362 (2003)
50. P. Wang, T. Ding, Creep of electrical resistance under uniaxial pressures for carbon black-silicone rubber composite. J. Mater. Sci. **45**, 3595–3601 (2010)

Chapter 3
Demonstrations and Applications of Fractional-Order Devices

3.1 Introduction

As discussed in Chap. 1, the need for broadband phase compensation in electronic circuits dates back to H. Bode's original text [5]. Bode considered achieving flat phase response so important that he abandoned two additional chapters in favor of completing a complex mix of resistors, capacitors, and inductors to meet his specifications. Ironically, just a few years earlier Cole and Cole discovered the power-law behavior in numerous materials and gave the characteristic the label constant phase element (CPE) [9]. Power-law CPE models are now a standard part of electrical impedance spectroscopy. Likewise, achieving flat phase response for phase compensation in control circuits is standard practice for linear as well as nonlinear systems. In this chapter, we continue with the challenge of putting these ideas together. We have shown that a single-component device can be constructed exhibiting nearly constant phase over a broad bandwidth. Now we address whether such devices can be integrated into control systems to achieve the performance improvements that have been predicted? Are these devices exhibiting "flat enough" phase response for useful applications?

3.2 Circuit and System Design Using Fractional-Order Elements

Looking again at the feedback operational amplifier circuit discussed in Chap. 1, the gain magnitude is determined by the ratio K/R_{IN} and the frequency scaling $1/\omega^\alpha$, and the phase is determined by $j^{-\alpha}$, or $\phi = -\pi/2$, which is not dependent on frequency. This is one of the main points for using such a device. You get a selectable phase shift that is maintained over a very broad bandwidth. This provides for an extensive array of options in designing robust phase margins for systems exhibiting large drifts in plant parameters.

The gain of a simple FO integrator implemented with an operational amplifier circuit shown in Fig. 3.1 is

© The Author(s) 2017
R. Caponetto et al., *Fractional-Order Devices*,
SpringerBriefs in Nonlinear Circuits, DOI 10.1007/978-3-319-54460-1_3

Fig. 3.1 Basic FO
integrator. The schematic
symbol for the fractor (Z_F)
is intended to give the
impression of an element
somewhere between a
resistor and capacitor

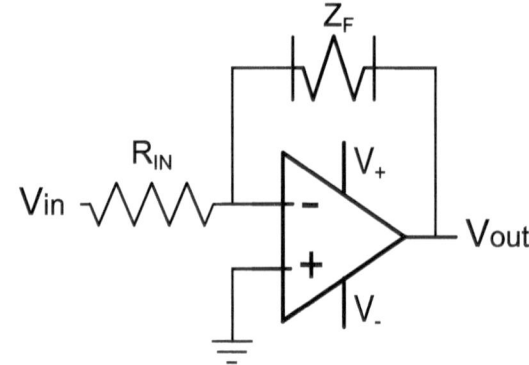

$$G = \frac{Z_F}{Z_{in}} = \frac{K}{R_{IN}} \frac{1}{(\tau s)^\alpha}. \tag{3.1}$$

Note that swapping the resistor and fractor positions would result in a FO differentiator. FO devices can be used exactly the same way conventional devices are employed.

3.3   Control System Demonstrations

A number of demonstrations were performed to verify the predictions. These demonstrations include temperature and robotic arm motion control applications in signal processing and sensing.

3.3.1   Temperature Control Demonstration

Theory says that you want a controller to match the order of the system. Thermal systems are diffusive with order 1/2, so a FO controller (FOC) of 1/2 order should be optimal [6]. A fractor of order approximately $\alpha \approx 0.5$ was inserted in the control circuit of a Wavelength Electronics^{TM} MPT-5000 to create a PI$^\alpha$ controller. The system under control was thermoelectric cooler (Peltier) plate attached to an aluminum plate of approximately 400 grams. No other tuning was required to achieve the change in response from Fig. 3.2a, b.

As can be seen, the settling occurred much more rapidly with FOC with hardly any detectable overshoot or undershoot of the setpoint.

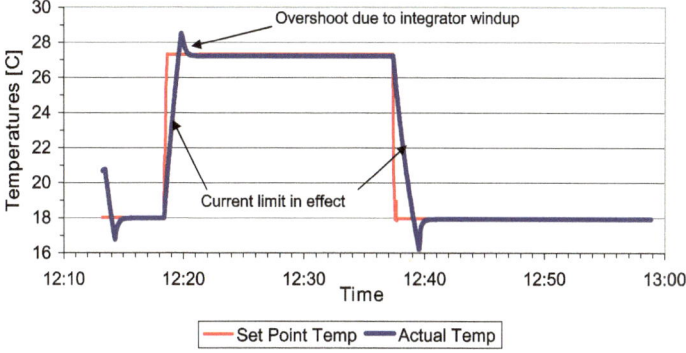

(a) The optimal conventional control (integer order) response.

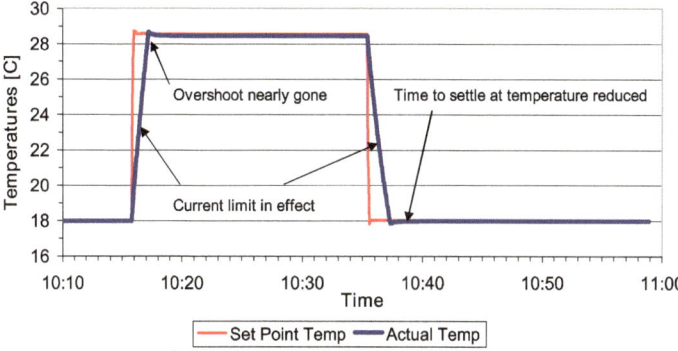

(b) The response achieved by swapping out the integrating capacitor with a prototype fractor. No other tuning was required.

Fig. 3.2 Overshooting or undershooting temperature in, for example, a diode laser system can result in destabilizing the laser or causing condensation which can adversely affect the operation and shorten the lifetime of the laser. Since many diode lasers can be tuned in frequency through temperature control, it is highly desirable to achieve the fastest possible settling over the widest possible frequency range while avoiding overshoot or undershoot of the target temperature [7]

3.3.2 Robotics Control Demonstration

To demonstrate the fractor in a robotics control application, we chose the rotary flexible joint (RFJ) from Quanser [14]. This system exhibited nonlinear sticky friction with an characteristic order of approximately 0.3. This system is considered as an advanced challenge due to the fact that the arm is only attached to the motorized hub passively through a set of springs as shown in Fig. 3.3. The FOC instrument in Fig. 3.3b was given the name "Fractroller".

If there were no nonlinear sticky friction or gear backlash in the RFJ challenge, straight proportional control would have been quite adequate. The suggested control solution includes state feedback from angle encoders in the hub (Θ) and arm (Φ)

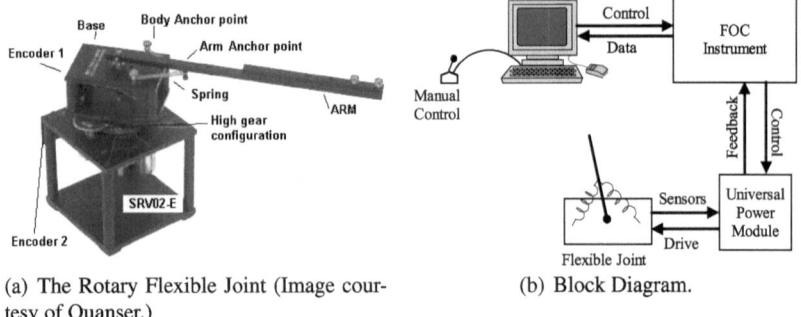

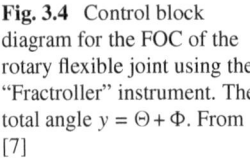

(a) The Rotary Flexible Joint (Image courtesy of Quanser.)

(b) Block Diagram.

Fig. 3.3 The rotary flexible joint and the experimental setup used for demonstrating mechatronic control. From Quanser by permission [14]

Fig. 3.4 Control block diagram for the FOC of the rotary flexible joint using the "Fractroller" instrument. The total angle $y = \Theta + \Phi$. From [7]

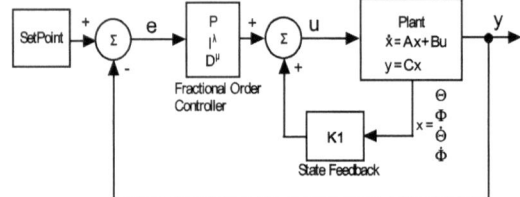

joints, as well as time derivatives of these signals in addition to the regular error signal feedback from the total angle ($\Theta+\Phi$) as shown in Fig. 3.4.

As can be seen in Fig. 3.5, proportional alone undershoots the setpoint and sticks. Proportional plus integral control overshoots the setpoint and sticks on the other side. Interestingly, the error signal builds up enough to cause the system to jump back past the set point after about 5 to 10 s, then sticks, then jumps again. This delayed oscillation about setpoint with PI control was observed for up to 30 min.

After some experimenting, we disabled the state feedback and implemented a simple PI^α controller with the result shown in Fig. 3.5c. With FOC, the setpoint was reached in less time than with any conventional control with negligible overshoot.

Attempts were made to cause the FO controller to become unstable by physically holding the arm offset from the setpoint for several seconds and then releasing it, expecting it to overshoot and go into oscillation prior to settling. This did not happen. It simple moved back to setpoint without overshoot or oscillation. Additionally, the arm configuration was changed to modify the moment of inertia of the arm. This would have required additional tuning of a conventional controller. With the fractional-order controller, it settled as if nothing had changed.

A mode of operation with no conventional equivalent involves setting the P term gain to zero and using only the I^α term in the controller. The result is a smooth, high efficiency "transport" motion. This is shown in Fig. 3.6. How many other new modes with no conventional equivalent exist is an open question.

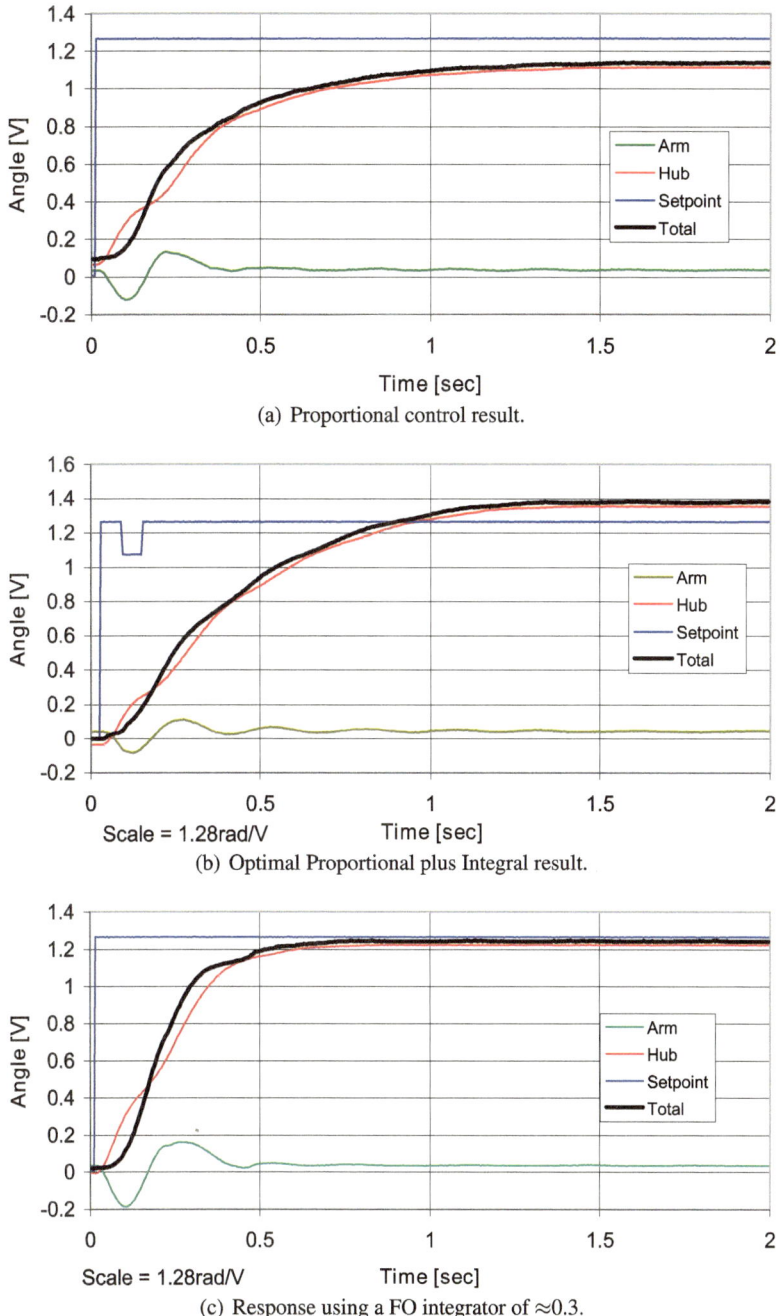

(a) Proportional control result.

(b) Optimal Proportional plus Integral result.

(c) Response using a FO integrator of ≈0.3.

Fig. 3.5 Comparison of traditional proportional (P), and conventional proportional plus integral (PI) control results versus PI$^\alpha$ control techniques [7]

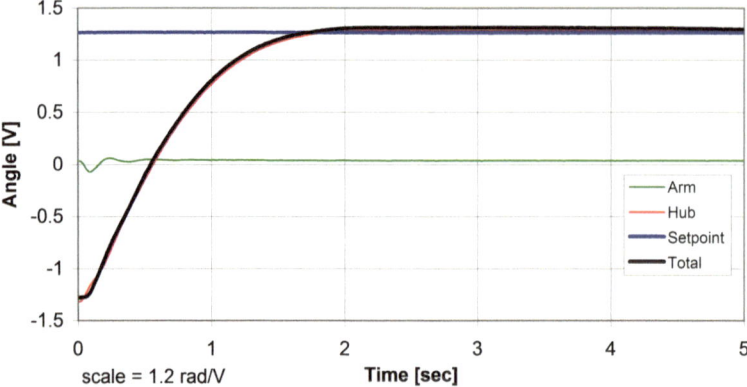

Fig. 3.6 Recorded operation of the test plant with the fractroller configured for FOC with the I^α term only

3.4   Cascaded Circuit Demonstration

One of the classic equations of physics is the harmonic oscillator. The model has been the subject of FO analysis as well [13]. Since this has a known solution, we chose to implement the solution as a feedback circuit to demonstrate the application of cascaded orders for FO operators [8].

$$\rho y^{(\alpha)} + \kappa y = f(t), \tag{3.2}$$

where $y^{(\alpha)}$ is shorthand for the FO derivative $_{-\infty}D_t^\alpha y$ and $f(t)$ being the forcing function. The solution can be formally expressed as

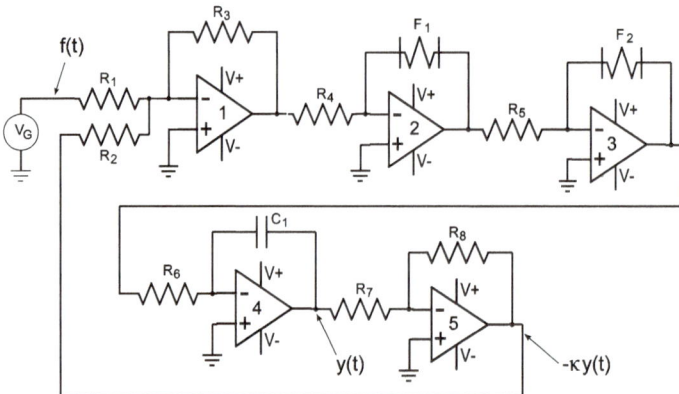

Fig. 3.7   Circuit schematic for fractional harmonic oscillator [8]

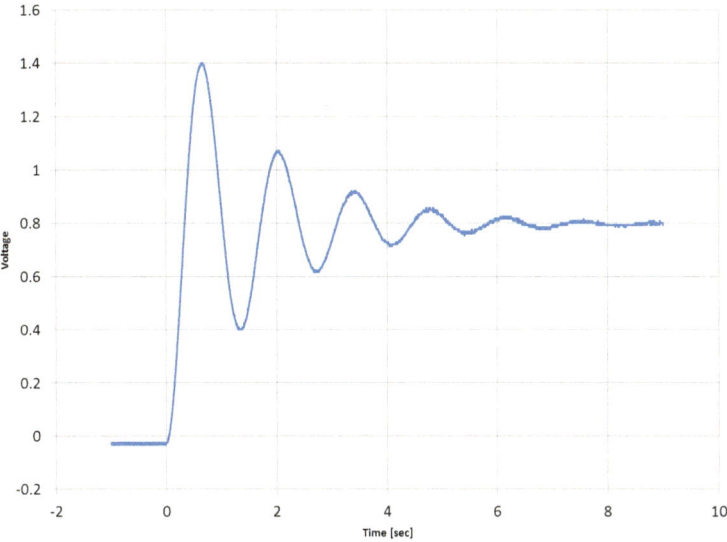

Fig. 3.8 Oscilloscope trace for FHO subjected to square wave input [8]

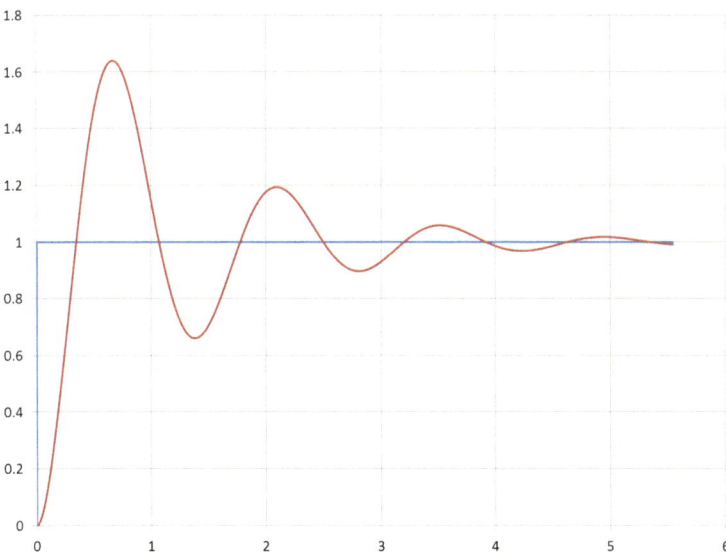

Fig. 3.9 Numerical simulation of solution of FO harmonic oscillator subjected to step input function [8]

$$y(t) = (1/\rho) \, _{-\infty}I_t^\alpha \{f(t) - \kappa y(t)\}. \tag{3.3}$$

The solution in the case of a $f(t) = \Theta(t)$, the step function turning on at time $t = 0$, can be written as

$$y(t) = \left(^1/_\rho\right) t^\alpha \sum_{n=0}^{\infty} \frac{(-1)^n (\frac{\kappa}{\rho} t^\alpha)^n}{\Gamma(n\alpha + 1 + \alpha)}. \tag{3.4}$$

The cascading of the amplifiers should result in the gains multiplying. This should result in addition of the exponents (the orders of the impedance). In the case shown in Fig. 3.7, fractors (described in Sect. 2.3.3) with orders of 0.3 and 0.5 were used in addition to a capacitor of order ≈ 1. The result was compared with the theoretical response of a FO harmonic oscillator of order 1.8. Comparing Figs. 3.8 and 3.9, we see that the model solution is played out in a physical circuit.

3.5 Circuit Demonstrations Using Solution-Based FOE

This section discusses the design and performance of simple integrator and differentiator (the building blocks of PID controller) circuits implemented using the solution-based FOE [3].

A FO differentiator circuit was constructed as shown in Fig. 3.10 as a variation of Fig. 3.1, but with the solution-based systems discussed in Chap. 2.

The transfer function (in the frequency domain where the phase is constant) for the fractional-order differentiator circuit can be written as

$$\frac{V_0}{V_i}(s) = -\frac{R_{Feedback}}{K}(\tau s)^\alpha \quad \text{or} \quad -R_{Feedback} C_F s^\alpha, \qquad 0 \le \alpha \le 1. \tag{3.5}$$

with C_F as described in Chap. 1.

Fig. 3.10 Fractional-order differentiator

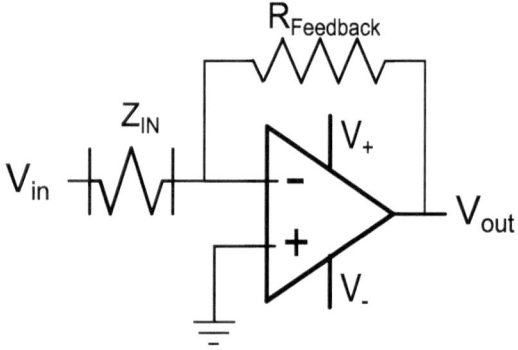

As an example of the differentiator operating on a triangular waveform, we can predict the performance by using the Laplace transform of the triangular wave of amplitude V as [10]

$$V_i(s) = \mathcal{L}V_i(t) = \frac{\int_0^T V_i(t)e^{-st}\,dt}{1 - e^{-st}} = \frac{4V}{Ts^2}\frac{(1 - e^{-sT/4})^2}{(1 + e^{-sT/2})}.$$

Substituting $V_i(s)$ in (3.5), we obtain

$$V_O(s) = -\frac{4RC_F s^\alpha V}{Ts^2}\frac{(1 - e^{-sT/4})^2}{(1 + e^{-sT/2})}. \tag{3.6}$$

Expanding and rearranging the numerator and denominator, $V_0(s)$ can be written as

$$V_0(s) = -\frac{4RC_F s^\alpha V}{Ts^2}[1 - 2e^{-sT/4} + 2e^{-3sT/4} - 2e^{-5sT/4} + \cdots].$$

Using the relation [11]

$$\mathcal{L}^{-1}(\frac{e^{-as}}{s^{1+n}}) = \frac{(t-a)^n}{n\Gamma(n)}u(t-a), \tag{3.7}$$

where "Γ" denotes the gamma function; the inverse Laplace transform of $V_0(s)$ is given by [3, 4]

$$V_0(t) = -\frac{4RC_F V}{T\,(1-\alpha)\,\Gamma(1-\alpha)}[t^{1-\alpha} - 2\sum_{n=0}^{\infty}(-1)^n$$

$$(t - \frac{2n+1}{4}T)^{1-\alpha}u(t - \frac{2n+1}{4}T)]. \tag{3.8}$$

The output $V_0(t)$ of the FO differentiator circuit is simulated using "Mathematica" taking V_i as a triangular input of amplitude 1 V peak. For simulation the series approximation up to $n = 25$ has been considered; there is no appreciable change for approximations beyond $n = 10$ (Figs. 3.11 and 3.12).

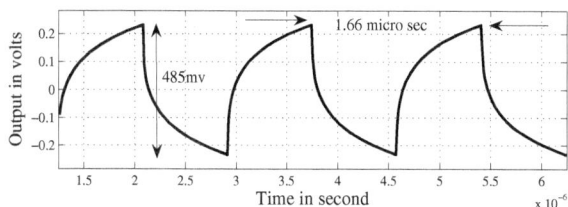

Fig. 3.11 Simulated output waveform of a FO differentiator at 600 kHz when $\alpha = 0.696$ and $C_F = 14.32 \times 10^{-9}$

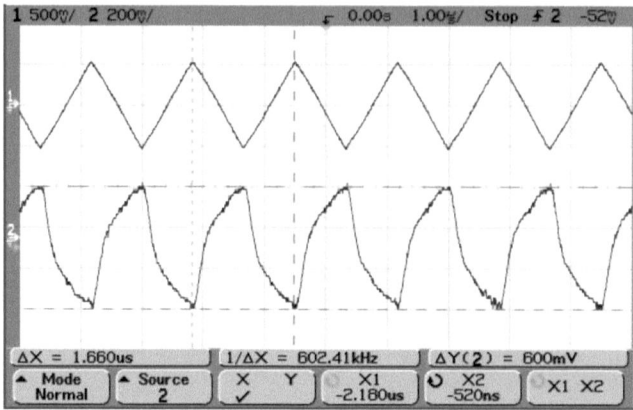

Fig. 3.12 Experimentally obtained output waveform of a FO differentiator at 600 kHz when $\alpha = 0.696$ and $C_F = 14.32 \times 10^{-9}$; data 1: input, data 2: output

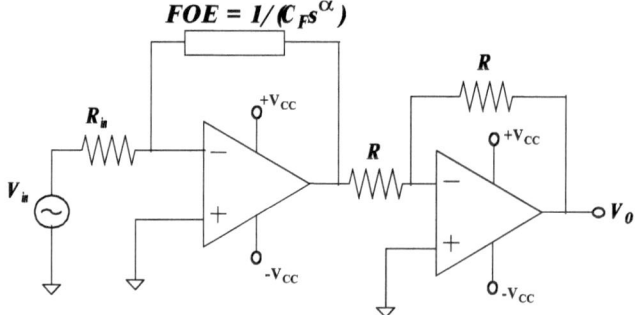

Fig. 3.13 FO Integrator with second operational amplifier to restore positive polarity

Similar to the differentiator circuit, the FO integrator has been implemented. The transfer function of FO integrator shown in Fig. 3.13 is given as

$$\frac{V_0(s)}{V_{in}(s)} = \frac{Q}{R_{in}s^\alpha} \tag{3.9}$$

or,

$$\frac{V_0(j\omega)}{V_{in}(j\omega)} = \frac{1}{R_{in}C_F(j\omega)^\alpha} = \frac{1}{R_{in}C_F\omega^\alpha} \times e^{-j\alpha\frac{\pi}{2}}. \tag{3.10}$$

Therefore,

$$Gain = \frac{1}{R_{in}C_F\omega^\alpha} \tag{3.11}$$

And,

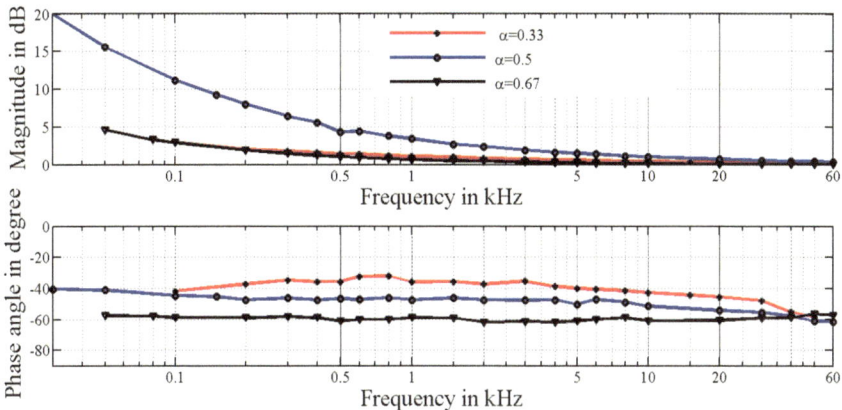

Fig. 3.14 Frequency response of FO integrator circuit

$$Phase = -\frac{\alpha\pi}{2} \tag{3.12}$$

The magnitude and phase response of the above circuit is shown in Fig. 3.14 for three different values of the exponent α.

3.5.1 Filter Circuit Demonstration

Sallen–Key Filter: The gain of the Sallen–Key filter is $K = 1 + R_4/R_3$ (refer Fig. 3.15). For an integer order filter $K > 3$, the filter becomes unstable but for a FO Sallen–Key filter, it is possible to design a stable filter for $K > 3$ (Table 3.1).

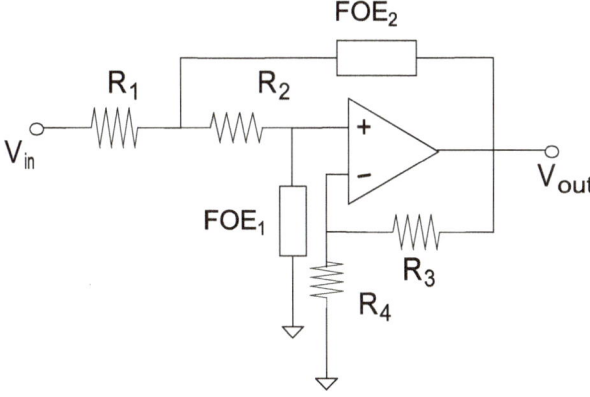

Fig. 3.15 FO Sallen–Key filter circuit

Table 3.1 The comparison of the simulated and experimental peak values of the FO differentiator output for 1 V peak input voltage; s: for simulated values and e: for experimentally obtained values

Frequency (kHz)	$\alpha = 0.503$	$\alpha = 0.538$	$\alpha = 0.617$	$\alpha = 0.696$
	$C_F = 10.55 \times 10^{-8}$ (mV)	$C_F = 7.14 \times 10^{-8}$ (mV)	$C_F = 3.67 \times 10^{-8}$ (mV)	$C_F = 14.32 \times 10^{-9}$ (mV)
200	s:120	s:139	s:182	s:243
	e:140	e:163	e:225	e:288
300	s:142	s:166	s:258	s:360
	e:184	e:209	e:288	e:381
400	s:166	s:192	s:298	s:362
	e:220	e:225	e:334	e:463
600	s:222	s:250	s:382	s:485
	e:292	e:284	e:431	e:600
800	s:252	s:283	s:442	s:580
	e:320	e:338	e:481	e:663

This can be explained from the transfer function of the filter

$$H(s) = \frac{K}{s^{\alpha+\beta}R^2C_F^2 + s^\alpha(1 - K)RC_F + 2RC_Fs^\beta + 1}. \tag{3.13}$$

For integer order case ($\alpha = \beta = 1$), the filter is unstable when $K > 3$ because the second coefficient of the characteristics equation, i.e., $(s^2R^2C_F^2 + s(3 - K)RC_F + 1)$ of the second-order Sallen–Key filter becomes negative. However, the filter can be stable in fractional domain, even when gain (K) is greater than 3. The condition of stability [15] for a FO filter ($K > 3$) is given as $p_1 < \frac{4\delta}{\pi}$, where $\delta = \frac{(K-3)a}{2\sqrt{b}} < \frac{\pi}{2}$, $a \le b$ and p_1 is the order of FO filter for the TF $T(s) = \frac{bK}{s^{\alpha+\beta}+s^\alpha(3-K)a+b}$. So, it is clear that the above characteristics equation with gain ($K = 4$) can be stable in fractional domain when $\alpha + \beta < 1.33$ and $b = a^2$ [17].

KHN biquad filter: Similarly the performance of the KHN biquad filter response can be studied in the fractional domain. The circuit diagram of a typical KHN biquad filter is shown in Fig. 3.16.

Considering $FOE_1 = 1/(C_{F1}s^\alpha)$ and $FOE_2 = 1/(C_{F2}s^\beta)$, the transfer function for the low-pass, high-pass, and band-pass can be expressed as

$$T_{lp} = \frac{\frac{2R_3}{R_2+R_3}}{(R_4R_5C_{F1}C_{F2})s^{\alpha+\beta} + \frac{2R_2}{R_2+R_3}(R_5C_{F2})s^\beta + 1}, \tag{3.14}$$

$$T_{hp} = \frac{(\frac{2R_3}{R_2+R_3})(R_4R_5C_{F1}C_{F2})s^{\alpha+\beta}}{(R_4R_5C_{F1}C_{F2})s^{\alpha+\beta} + \frac{2R_2}{R_2+R_3}(R_5C_{F2})s^\beta + 1}, \tag{3.15}$$

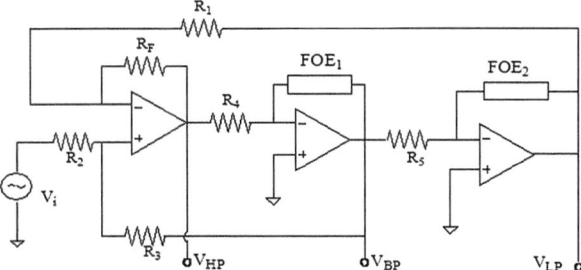

Fig. 3.16 FO KHN biquad filter circuit

$$T_{bp} = \frac{\frac{\frac{2R_3}{R_2+R_3}}{\frac{R_2+R_3}{2R_2}}(R_5C_{F2})s^\beta}{(R_4R_5C_{F1}C_{F2})s^{\alpha+\beta} + \frac{1}{(\frac{R_2+R_3}{2R_2})}(R_5C_{F2})s^\beta + 1}. \tag{3.16}$$

The analysis of the above equations shows that the extra parameters α and β provide greater flexibility in designing the peak frequency, cutoff frequency, and Q factor of the filter.

3.5.2 PLL Circuit Demonstration

The block diagram of a PLL circuit is shown in Fig. 3.17. An FO analog PLL can be designed in three ways:

- by using an FO low-pass filter,
- by using an FO VCO or,
- by using both FO low-pass filter and FO VCO.

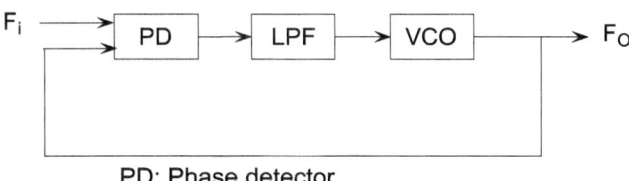

PD: Phase detector
LPF: Low pass filter
VCO: Voltage controlled oscillator

Fig. 3.17 Block diagram of phase-locked loop

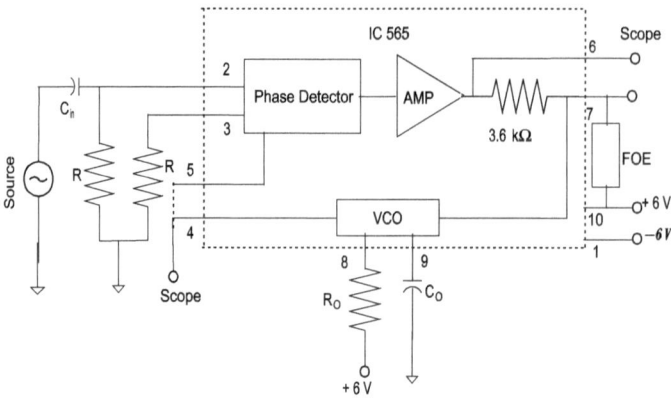

Fig. 3.18 FO PLL circuit

When both FO LPF and FO VCO are used in FPLL, TF that relates the phase error for FPLL can be obtained as

$$\frac{\theta_e(s)}{\theta_i(s)} = \frac{s^\beta (s^\alpha + b)}{s^{\alpha+\beta} + s^\beta b + K},\qquad(3.17)$$

where, $K = K_f K_d K_0$, K_0 is the VCO sensitivity, K_d is the sensitivity of phase detector (PD), K_f is the sensitivity of the loop filter, θ_i is the phase of reference or input signal, and θ_e is the output phase of phase detector. α and β are the orders of FO LF and FO VCO, respectively.

Similarly, the relation between input signal phase and VCO output phase can be expressed by the following transfer function.

$$\frac{\theta_0(s)}{\theta_i(s)} = \frac{K}{s^{\alpha+\beta} + s^\beta b + K},\qquad(3.18)$$

where θ_0 is the output phase of VCO. For $\alpha = \beta = 1$, the transfer functions (3.17) and (3.18) simplify to a second-order PLL.

Figure 3.18 shows the circuit diagram of an analog FPLL using IC565. The low-pass filter contains FOE in place of the capacitance to design the fractional PLL.

Figures 3.19 and 3.20 compare the transient response of the integer order PLL and the FO PLL. To observe the transient response of PLL, a sinusoidal signal of varying frequency within the lock range is applied. The incoming input signal of frequency 550 Hz is changed to 350 Hz with 180° phase difference at time 5 ms, and applied to both the integer order PLL and FO PLL. The TF of the PLL is described by (3.17) and (3.18). It will be worth to mention that for integer order PLL both α and β are 1. While for FO PLL, the value of $\alpha = 0.43$ as the low-pass filter contains the FOE with the exponent value of 0.43 [16].

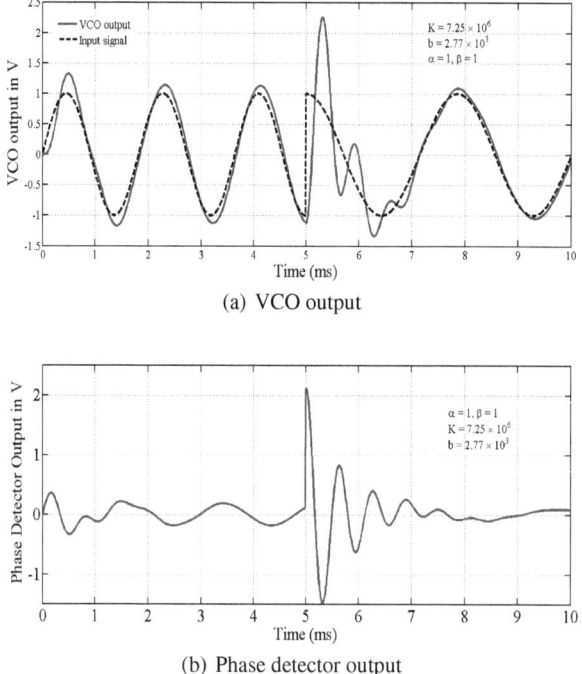

(a) VCO output

(b) Phase detector output

Fig. 3.19 Transient response of integer order PLL

From the above figures, it is apparent that the VCO tracks the same change in input signal at a faster pace without significant oscillations in case of fractional PLL circuit as the FPLL lock time is much less than that for IPLL [17].

3.5.3 Resonator Circuit Demonstration

Realization of inductor using FOE and GIC network

As per the knowledge of the authors, to date no single-component fractional-order inductor is available in literature [18]. However, a fractional-order inductor can be realized using a GIC network as shown in Fig. 3.21. It is apparent from the figure that if capacitor or FOE is used to replace the resistor then different FOEs with $-2 < \alpha < +2$ can be realized [1]. Such inductor can be used to design a resonator circuit with high Q factor [2] as shown in Fig. 3.22. Figure 3.23 shows the magnitude and phase response of the FO parallel resonator circuit. The experimentally obtained Q factor is about 85 which is much higher than the Q factor obtained by any integer order resonator circuit in this frequency range of operation [2].

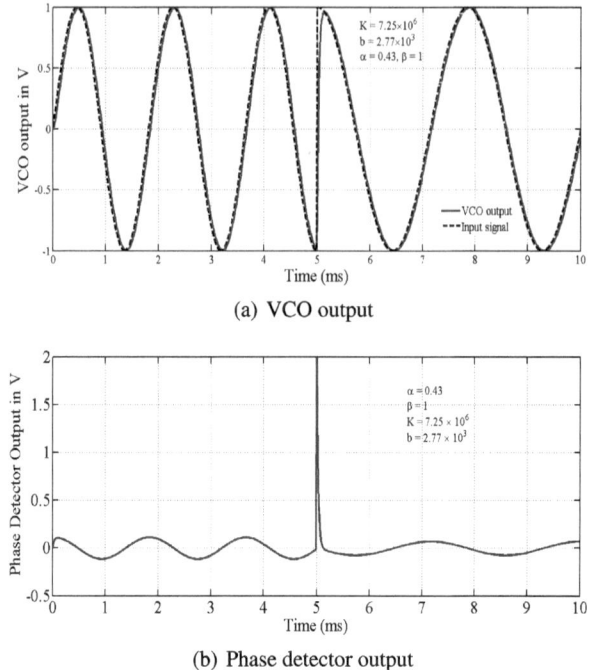

(a) VCO output

(b) Phase detector output

Fig. 3.20 Transient response of FO r PLL (FPLL)

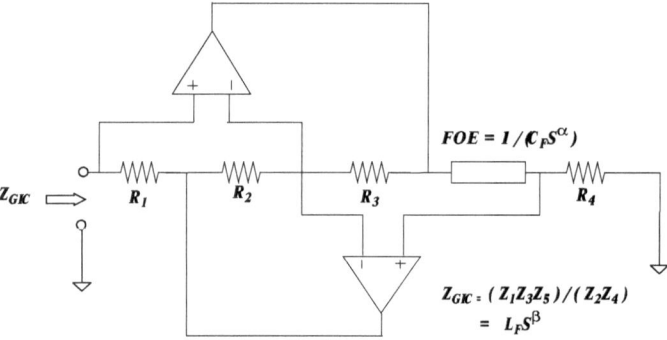

Fig. 3.21 FO inductor realized using FOE and GIC network

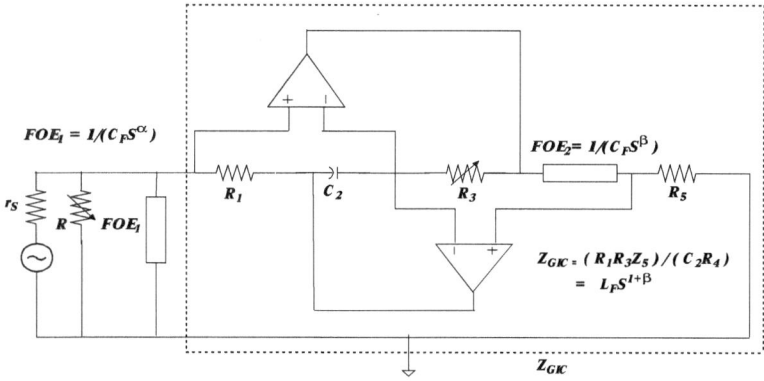

Fig. 3.22 FO parallel resonator circuit

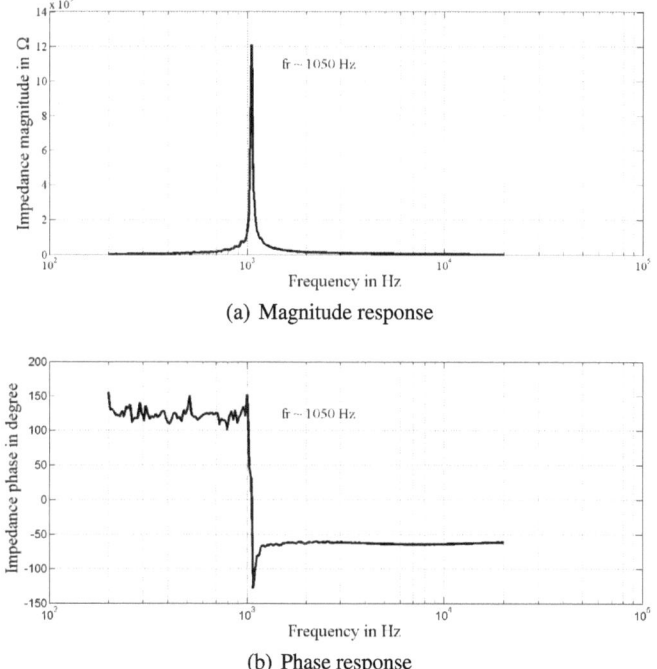

Fig. 3.23 Experimentally obtained magnitude and phase plot of FO parallel resonator circuit

3.6 Conclusion

At this point, it is clear that analog devices can be built and will respond as predicted. Such devices can provide bandwidth of response over seven decades or more of frequency, depending on the requirements of the problem at hand. Designing with

FOE is no more difficult that with any other element and, in fact, allows for the creation of much simpler circuits which have been demonstrated to substantially improve the performance. As important, the specific structure of the FOE is not critical. Demonstrations using radically different construction produced results well within the expectations of prototype devices. There is no shortage of additional problems to be addressed with fractional-order devices. For example, see [12] for an extensive discussion of applications of FOC systems.

References

1. A. Adhikary, P. Sen, S. Sen, K. Biswas, Design and performance study of dynamic fractors in any of the four quadrants. Circuits Syst. Signal Process. **35**(6), 1909–1932 (2015)
2. A. Adhikary, S. Sen, K. Biswas, Practical realization of tunable fractional order parallel resonator and fractional order filters. Trans. Circuits Syst. I **63**(8), 1142–1151 (2016)
3. K. Biswas, S. Sen, P.K. Dutta, Realization of a constant phase element and its performance study in a differentiator circuit. IEEE Trans. Circuits Syst. **II**(53), 802–806 (2005)
4. K. Biswas, Studies on design, development and performance analysis of capacitive type sensors. Ph.D. Thesis, Indian Institute of Technology Kharagpur, India, Department of Electrical Engineering–India (2007)
5. H. Bode, *Network Analysis and Feedback Amplifier Design* (Van Nostrand, New York, 1945)
6. G.W. Bohannan, Analog fractional order controller in a temperature control application, in *2nd IFAC Workshop on Fractional Differentiation and its Applications, FDA'06* (2006), pp. 40–45
7. G.W. Bohannan, Analog fractional order controller in temperature and motor control applications. J. Vib. Control **14**(9–10), 1487–1498 (2008)
8. G.W. Bohannan, B. Knauber, A physical experimental study of the fractional harmonic oscillator, in, International Symposium on Circuits and Systems (ISCAS'15), 24–27 May 2015. Lisbon **2015**, 2341–2344 (2015)
9. S. Cole, R. Cole, Dispersion and absorption in dielectrics. J. Chem. Phys. **9**, 341–351 (1941)
10. R.A. Gabel, R.A. Roberts, *Signals and Linear Systems*, 3rd edn. (Wiley, Singapore, 1995)
11. E. Kreyszig, *Advanced Engineering Mathematics* (Wiley, Singapore, 1999)
12. C.A. Monje, Y.Q. Chen, B.M. Vinagre, D. Xue, V. Feliu, *Fractional-Order Systems and Controls: Fundamentals and Applications* (A Monograph in the Advances in Industrial Control Series (Springer, Berlin, 2010)
13. I. Podlubny, T. Skovarnek, B.M. Vinagre, V. Veritsky, Y.Q. Chen, Matrix approach to discrete fractional calculus III: non-equidistant grids, variable step length and distributed orders. Phil. Trans. R. Soc. **371**(20120153), 1–15 (2013)
14. Quanser, Rotary Flexible Joint, Markham, Ontario, product information, http://www.quanser.com/Products/rotaryflexiblejoint
15. A.G. Radwan, A.M. Soliman, A.S. Elwakil, On the stability of linear system with fractional order elements. Chaos, Solitons Fractals **40**, 2317–2328 (2009)
16. M.C. Tripathy, D. Mondal, K. Biswas, S. Sen, Design and performance study of phaselocked loop using fractional order loop filter. Int. J. Circuit Theory Appl. 43, 776–792 (2015)
17. M.C. Tripathy, Design and performance analysis of fractional order filters. Ph.D. Thesis, Indian Institute of Technology Kharagpur, India, Department of Electrical Engineering–India (2015)
18. J.A. Tenreiro Machado, M.S.G.F. Alexandra, Fractional order inductive phenomena based on the skin effect. Nonlinear Dyn. **68**, 107 (2012). doi:10.1007/s11071-011-0207-z

Chapter 4
Fractional-Order Models of Vegetable Tissues

4.1 Introduction

Biological systems have a hierarchical and fractal structure with many levels at different scales, ranging from organs down to tissues and cells, that are difficult to model by classical mathematical tools. Fractional-order (FO) models are more adequate to describe living matter. In particular, vegetable tissues can be thought of as nature's devices exhibiting FO dynamics over a very wide frequency ranges.

We investigate living tissues from the perspective of the electrochemist and electrical engineer by means of electrical impedance spectroscopy (EIS). This exploration can help making additional connections between and among many different, and often divergent, specialties in biology, chemistry, physics, and engineering.

EIS measures the electrical impedance of a specimen within a given range of frequencies, producing a spectrum that represents the variation of the impedance versus frequency [25, 31–33]. EIS has been widely used for studying vegetable tissues, namely roots, leaves, stems, fruits, and vegetables. Table 4.1 lists several references on the topic.

On reviewing the discussion of impedance spectroscopy, EIS involves exciting a specimen with frequency-variable electric sinusoidal signals and registering the system response. The voltage $v(t)$ and current $i(t)$ across the specimen at steady state are sinusoidal functions of time given by

$$v(t) = V \cos(\omega t + \theta_V), \tag{4.1a}$$

$$i(t) = I \cos(\omega t + \theta_I), \tag{4.1b}$$

where $\{V, I\}$ are the amplitudes of the voltage and current, $\{\theta_V, \theta_I\}$ denote their phase shifts, and ω represents the angular frequency.

The voltage and current can be represented in the frequency domain by

$$\mathbf{V}(j\omega) = V \cdot e^{j\theta_V}, \tag{4.2a}$$

$$\mathbf{I}(j\omega) = I \cdot e^{j\theta_I}, \tag{4.2b}$$

© The Author(s) 2017
R. Caponetto et al., *Fractional-Order Devices*,
SpringerBriefs in Nonlinear Circuits, DOI 10.1007/978-3-319-54460-1_4

Table 4.1 Several references reporting impedance models for vegetable tissues

Type of tissue	Main objective	Reference
Roots	Estimate size	[5]
Roots	Monitoring growth	[46]
Fruits and vegetables	Food quality control	[12]
Fruits and vegetables	Physical characteristics	[25]
Shoots and leaves	Seasonal characteristics	[34]
Vegetables	Industrial processing	[58]
Wood	Disease investigation	[52]
Leaves	Physical characteristics	[31]
Vegetables	Physical characteristics	[32]
Vegetables	Freezing damage	[62]
Fruits	Effect of bruising and pressure	[21]
Leaves	Growth assessment	[37]
Leaves	Environment stresses	[60]
Roots	Absorptive root surface area	[54]
Roots	Estimate size	[13]
Leaves	Physical characteristics	[63]
Leaves	Disease investigation	[22]
Leaves	Physical characteristics	[61]
Leaves	Frost damage	[48]
Vegetables	Behavior during drying	[2]
Stems and leaves	Molecular structure	[28]
Fruits and vegetables	Assessing moisture contents	[27]
Fruits and vegetables	Physical characteristics	[40]
Vegetables	Industrial processing	[59]
Vegetables	Industrial processing	[43]
Fruits	Industrial processing	[53]
Vegetables	Industrial processing	[11]
Diverse food	Food processing review	[47]
Fruits	Indirect measurement method	[35]
Fruits	Indirect measurement method	[17]
Diverse food	Assessing environmental effects	[39]
Fruits	Tissue anatomy and physiology	[6]
Vegetables	Effects of dehydrofreezing	[1]
Vegetables	Effects of drying	[57]
Leaves	Frost hardiness	[55]
Leaves	Freezing effects	[26]
Plants	Disease investigation	[3]
Diverse biological materials	Survey	[16]

where $j = \sqrt{-1}$. The complex impedance $\mathbf{Z}(j\omega)$ is defined as the ratio of phasors:

$$\mathbf{Z}(j\omega) = \frac{\mathbf{V}(j\omega)}{\mathbf{I}(j\omega)} = \frac{V}{I} \cdot e^{j \arg(\theta_V - \theta_I)} = |\mathbf{Z}(j\omega)| \cdot e^{j \arg[\mathbf{Z}(j\omega)]}. \tag{4.3}$$

Given an impedance spectrum obtained from experimental data, we need to find a mathematical description, that is, a "model" that fits well into the numerical values and has limited number of parameters [27, 36, 40]. Different empirical models developed in the context of the so-called dielectric relaxation phenomenon were used for that purpose [16].

Extracting the intrinsic properties of the material from the measured impedance is then achieved by

$$\varepsilon = \varepsilon' + j\varepsilon'' = \frac{1}{j\omega C_A Z}, \tag{4.4}$$

where C_A is the capacitance of the empty test cell used in the measurements.

4.2 Empirical Fractional-Order Models

The Cole–Cole (CC), Cole–Davidson (CD), and Havriliak–Negami (HN) models are anomalous relaxation models that generalize the Debye (D) equation [7, 8, 10, 14, 23, 29, 51, 56]. Even though the original models are empirical and do not use fractional derivatives or integrals, explicitly, they may be regarded as pioneer applications of fractional calculus. Such connection has been addressed by several authors [19, 24, 41, 42, 44, 49, 50].

The D, CC, CD, and HN models, in the Laplace domain, can be written as follows [49]:

$$\tilde{\varepsilon}_D(s) = \frac{\varepsilon^*(s) - \varepsilon_\infty}{\varepsilon_0 - \varepsilon_\infty} = \frac{1}{1 + s\tau}, \tag{4.5}$$

$$\tilde{\varepsilon}_{CC}(s) = \frac{1}{1 + (s\tau)^\alpha}, \quad 0 < \alpha \leq 1, \tag{4.6}$$

$$\tilde{\varepsilon}_{CD}(s) = \frac{1}{(1 + s\tau)^\beta}, \quad 0 < \beta \leq 1, \tag{4.7}$$

$$\tilde{\varepsilon}_{HN}(s) = \frac{1}{[1 + (s\tau)^\alpha]^\beta}, \quad 0 < \alpha \leq 1, \ 0 < \beta \leq 1, \tag{4.8}$$

where s is the Laplace variable, $\tilde{\varepsilon}$ denotes the complex susceptibility, $\{\varepsilon_0, \varepsilon_\infty\}$ represent the low- and high-frequency limits of the complex dielectric permittivity, ε^*, constant τ is the relaxation time, and $f = \omega/2\pi$ denotes frequency.

For some values of the exponents, the HN model yields the D, CC, and CD counterparts.

The D, CC, CD, and HK functions are solutions of their kinetic equations [49], and the relation between the complex dielectric permittivity, $\varepsilon^*(j\omega)$, and the relaxation function, $\phi(t)$, is given by

$$\tilde{\varepsilon}(j\omega) = \frac{\varepsilon^*(j\omega) - \varepsilon_\infty}{\varepsilon_0 - \varepsilon_\infty} = \mathscr{F}\left\{-\frac{d\phi(t)}{dt}\right\}, \tag{4.9}$$

where t is the time, $\mathscr{F}\{f(t)\} = \tilde{f}(j\omega)$ denotes the Fourier transform of $f(t)$ in the variable $j\omega$, and $\phi(0) = 1$.

Knowing that $\phi(t)$ obeys the expression,

$$\frac{d\phi(t)}{dt} = -\frac{d}{dt}\int_0^t M(t-x)\phi(x)dx, \tag{4.10}$$

where $M(t)$ represents the so-called integral memory function, and the kinetic equations associated with the D, CC, CD, and HN models are, respectively [49],

$$\frac{d\phi(t)}{dt} + \frac{1}{\tau}\phi(t) = 0, \tag{4.11}$$

$$\frac{d\phi(t)}{dt} + \frac{1}{\tau^\alpha}D_t^{1-\alpha}\phi(t) = 0, \tag{4.12}$$

$$\frac{d\phi(t)}{dt} + \frac{1}{\tau^\beta}\frac{d}{dt}\left\{e^{-t/\tau}\int_0^t (t-x)^{\beta-1}E_{\beta,\beta}^1\left[\left(\frac{t-x}{\tau}\right)^\beta\right]e^{x/\tau}\phi(x)dx\right\} = 0, \tag{4.13}$$

$$\frac{d}{dt}\left\{\phi(t) + \sum_{k=0}^\infty \int_0^t \frac{(t-x)^{\alpha\beta(k+1)-1}}{\tau^{\alpha\beta(k+1)}}E_{\alpha,\alpha\beta(k+1)}^{\beta(k+1)}\left[-\left(\frac{t-x}{\tau}\right)^\alpha\right]\phi(x)dx\right\} = 0. \tag{4.14}$$

The solutions of these equations are given by [49]

$$\phi_D(t) = e^{-t/\tau}, \tag{4.15}$$

$$\phi_{CC}(t) = E_{\alpha,1}\left[-\left(\frac{t}{\tau}\right)^{\alpha}\right], \tag{4.16}$$

$$\phi_{CD}(t) = 1 - \left(\frac{t}{\tau}\right)^{\beta} E_{1,\beta+1}^{\beta}\left[-\frac{t}{\tau}\right], \tag{4.17}$$

$$\phi_{HN}(t) = 1 - \left(\frac{t}{\tau}\right)^{\alpha\beta} E_{\alpha,\alpha\beta+1}^{\beta}\left[-\left(\frac{t}{\tau}\right)^{\alpha}\right], \tag{4.18}$$

where $D_t^{\gamma} f(t)$ represents the Riemann–Liouville fractional derivative and $E_a(\cdot)$, $E_{a,b}(\cdot)$, and $E_{a,b}^c(\cdot)$ denote the one-, two-, and three-parameter Mittag-Leffler functions, defined by

$$E_a(x) = \sum_{k=0}^{\infty} \frac{x^k}{\Gamma(ak+1)}, \tag{4.19}$$

$$E_{a,b}(x) = \sum_{k=0}^{\infty} \frac{x^k}{\Gamma(ak+b)}, \tag{4.20}$$

$$E_{a,b}^c(x) = \sum_{k=0}^{\infty} \frac{(c)_k}{\Gamma(ak+b)k!} x^k, \tag{4.21}$$

where the parameters $\{a, b, c\} \in \mathbb{C}$, with $\mathrm{Re}(\{a, b, c\}) > 0$, and expression $(c)_k = \frac{\Gamma(c+k)}{\Gamma(c)}$ represents the Pochhammer symbol.

Alternatively [41], defining the operator $[1 + (\omega_p)^{-\alpha}(_0D_t)^{\alpha}]^{\beta}$, where $\omega_p = \frac{1}{\tau}$, we obtain the HN model:

$$[1 + (\omega_p)^{-\alpha}(_0D_t)^{\alpha}]^{\beta} f(t) = W_{\omega_p}^{\alpha,\beta} f(t) = \tag{4.22}$$
$$\omega_p^{-\alpha\beta} \exp\left(-\frac{\omega_p^{\alpha} t}{\alpha} {_0D_t^{1-\alpha}}\right) \cdot {_0D_t^{\alpha\beta}} \exp\left(\frac{\omega_p^{\alpha} t}{\alpha} {_0D_t^{1-\alpha}}\right) f(t).$$

The D, CC, and CD equations can be obtained for some values of (α, β).

Other empirical models found in the literature are the Jurlewicz–Weron–Stanislavsky (JWS), or modified HN, the Kohlrausch–Williams–Watts (KWW), and the Capelas–Mainardi–Vaz (CMV) models. In the Laplace domain, those models can be expressed by [20, 44, 45]

$$\tilde{\varepsilon}_{JWS}(s) = 1 - \frac{1}{[1 + (s\tau)^{-\alpha}]^{\beta}} = 1 - (s\tau)^{\alpha\beta} \cdot \tilde{\varepsilon}_{HN}(s), \ 0 < \alpha \le 1, \ 0 < \beta \le 1,$$
(4.23)

$$\tilde{\varepsilon}_{KWW}(s) = \mathscr{L}\left\{ \frac{\beta}{\tau}\left(\frac{t}{\tau}\right)^{\beta-1} \cdot \exp\left[-\left(\frac{t}{\tau}\right)^{\beta}\right] \right\}, \ 0 < \beta \le 1,$$
(4.24)

$$\tilde{\varepsilon}_{CMV}(s) = \frac{1}{s}\left[1 + \sum_{n=1}^{\infty}(-1)^n \frac{\Gamma(n\beta + 1)}{s^{n\beta}} \prod_{i=0}^{n-1} \frac{\Gamma(i\beta + \beta - \alpha + 1)}{\Gamma(i\beta + \beta + 1)} \right],$$
(4.25)

$$0 < \alpha \le 1, \ 0 < \beta \le 1,$$
(4.26)

where their corresponding relaxation functions, $\phi(t)$, are

$$\phi_{JWS}(t) = E_{\alpha,1}^{\beta}\left[-\left(\frac{t}{\tau}\right)^{\alpha} \right],$$
(4.27)

$$\phi_{KWW}(t) = \exp\left[-\left(\frac{t}{\tau}\right)^{\beta} \right],$$
(4.28)

$$\phi_{CMV}(t) = E_{\alpha,\frac{\beta}{\alpha},\frac{\beta-\alpha}{\alpha}}(-t^{\beta})$$
(4.29)

$$= 1 + \sum_{n=1}^{\infty}(-1)^n \prod_{i=0}^{n-1} \frac{\Gamma(i\beta + \beta - \alpha + 1)}{\Gamma(i\beta + \beta + 1)} t^{\beta n}.$$
(4.30)

For particular values of the parameters, the CMV recovers the D ($\alpha = \beta = 1$), CC ($\alpha = 1$), and KWW ($\beta = 1$) models.

Biological tissues involve complex phenomena characterized by dynamic processes that occur at different lengths and timescales. Therefore, we verify that a simple D relaxation model cannot describe adequately the behavior of such materials, as it is unable to consider the interactions among various relaxing phenomena and memory effects [15]. The non-integer formulations of the D model are necessary to represent accurately the dynamical phenomena that take place within the material at different scales [4, 9, 18, 48, 56, 61].

4.3 On the Fractional-Order Models of Vegetable Tissues

In this section, we present two examples of EIS applied to vegetable tissues. The impedance spectra of the samples are measured and empirical models are fitted into the experimental data.

4.4 Modeling Different Size Stems of a Plant

In this first example, six different size stems, $\{S_1, \ldots, S_6\}$, of a *Hibiscus* plant (https:// en.wikipedia.org/wiki/Hibiscus) are analyzed. For the experiments, we used the setup depicted in Fig. 4.1 [31–33]. The stems were immersed in salted water at room temperature (25 °C), except their base. Two 0.5-mm-diameter copper electrodes were used to connect the specimens to the measurement equipment. One 10-mm electrode was inserted into the stem base, aligned with its main longitudinal axis. The other electrode was placed immersed in the salted water. Electrodes with different sizes were tested, but their influence on the results was found to be negligible. An adaptation resistance, $R_s = 15$ kΩ, in series with the stem under analysis was used. The electrical impedance was measured in the frequency range from $f = 10$ Hz to $f = 1$ MHz, at $L = 30$ logarithmically spaced points.

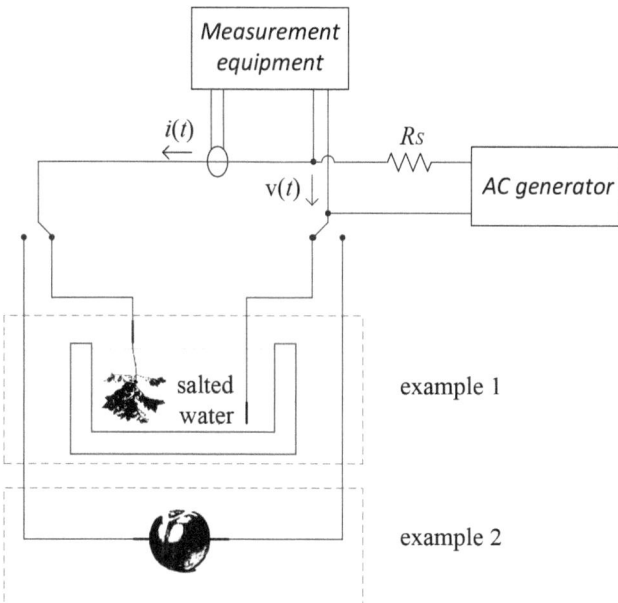

Fig. 4.1 Experimental setup for measuring impedance

The parameters of the FO empirical models were determined in order to minimize the Canberra-based distance, J, between the experimental, $\mathbf{Z}_e$, and model, $\mathbf{Z}_m$, impedances, that is,

$$J = \frac{1}{L} \sum_{k=1}^{L} \cdot \left(\frac{|\text{Re}[\mathbf{Z}_e(j\omega_k)] - \text{Re}[\mathbf{Z}_m(j\omega_k)]|}{|\text{Re}[\mathbf{Z}_e(j\omega_k)]| + |\text{Re}[\mathbf{Z}_m(j\omega_k)]|} \right) + \tag{4.31}$$
$$\left(\frac{|\text{Im}[\mathbf{Z}_e(j\omega_k)] - \text{Im}[\mathbf{Z}_m(j\omega_k)]|}{|\text{Im}[\mathbf{Z}_e(j\omega_k)]| + |\text{Im}[\mathbf{Z}_m(j\omega_k)]|} \right),$$

where $\text{Re}(\cdot)$ and $\text{Im}(\cdot)$ represent the real and imaginary parts of the argument.

The fitness function (4.31) captures the relative error of the adjustment, avoiding "saturation" effects, caused by the simultaneous presence of large and small values, that occur when using the standard Euclidean norm.

The experiments demonstrated that the best fit occurs for the six-parameter model:

$$\mathbf{Z}_{mod}(j\omega) = K \cdot \frac{\left(1 + \frac{j\omega}{z_1}\right)^{\alpha_1} \cdot \left(1 + \frac{j\omega}{z_2}\right)^{\alpha_2}}{(j\omega)^{\beta}}, \tag{4.32}$$

with $\{K, \beta, \alpha_1, z_1, \alpha_2, z_2\} > 0$.

The heuristic expression (4.32) represents a good compromise between model complexity and quality of fitting between experimental and analytical results.

Table 4.2 summarizes the parameters obtained for the six stem sizes analyzed. The experimental and approximated curves are depicted in Figs. 4.2, 4.3, and 4.4 in the form of Bode, Nichols, and polar plots.

Figure 4.5 compares the six spectra, where we observe a clear relationship between the stem size and the loci of $\mathbf{Z}(j\omega)$.

4.5 Modeling Fruits and Vegetables

Here, a second example is considered. We analyze three fruits, namely grape, plum, and persimmon $\{GR, PL, PE\}$, and three vegetables, namely pumpkin, courgette, and French garlic $\{PU, CO, FG\}$. For the experiments, the electrodes connecting the specimens to the measurement equipment were inserted into opposite sides of the fruit or vegetable, aligned with one main axis of symmetry (Fig. 4.1). Therefore, in this case the samples are not immersed in salted water.

Table 4.3 summarizes the parameters obtained for the six specimens of fruits and vegetables when fitting the empirical model of (4.32). The experimental and approximated curves representing the spectra are depicted in Fig. 4.6 in the form of Bode diagrams. The polar and Nichols plots are omitted for the sake of parsimony.

Table 4.2 Fractional-order model parameters of the *Hibiscus* stems in example 1

Stem sample	Shape	K ($\times 10^3$)	β	z_1 ($\times 10^5$)	α_1	z_2 ($\times 10^3$)	α_2	J
S_1		9.3	0.06	3.9	0.85	8.5	0.085	0.1579
S_2		11.5	0.04	3.5	0.95	8.5	0.200	0.1400
S_3		8.5	0.08	3.4	0.85	10.5	0.085	0.1444
S_4		9.2	0.06	5.5	0.80	4.0	0.080	0.1575
S_5		9.2	0.09	3.7	0.8	9.0	0.080	0.1334
S_6		11.0	0.08	2.8	0.70	10.0	0.220	0.1638

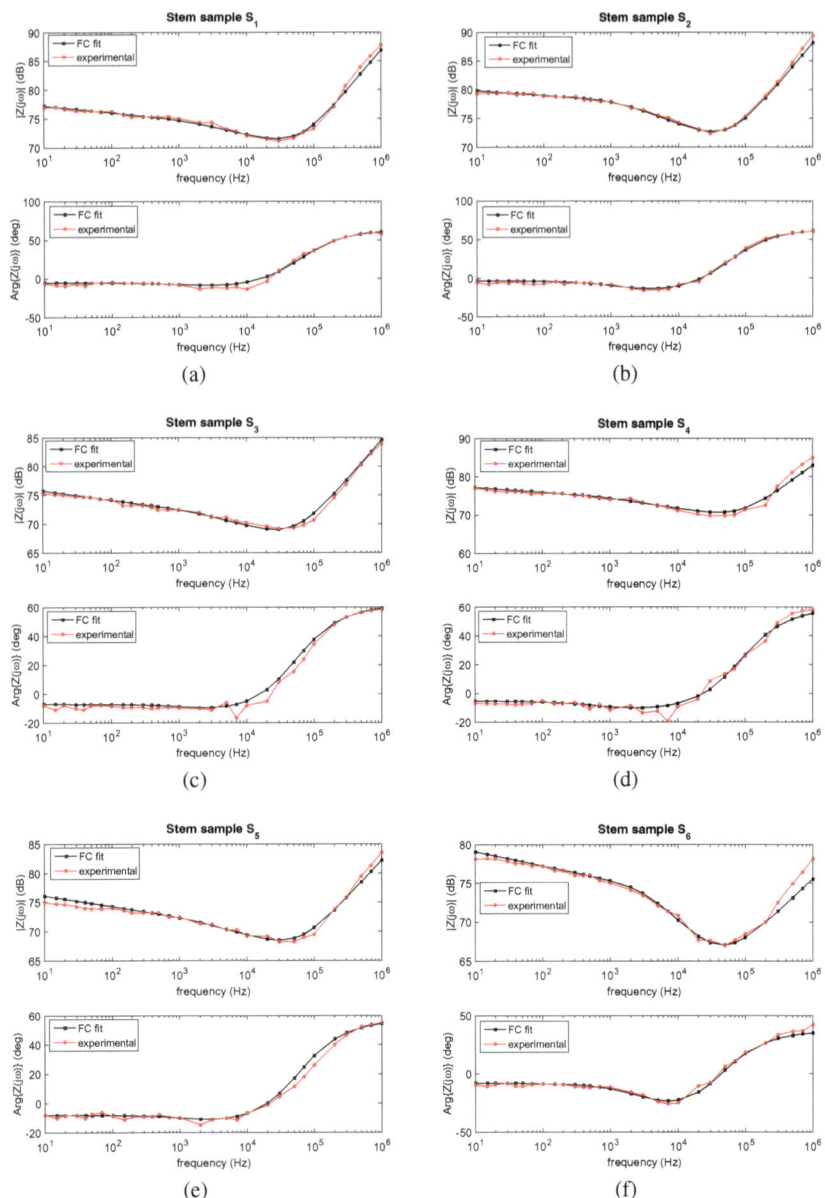

Fig. 4.2 Bode plots for the impedance $\mathbf{Z}(j\omega)$ of the six *Hibiscus* stems in example 1

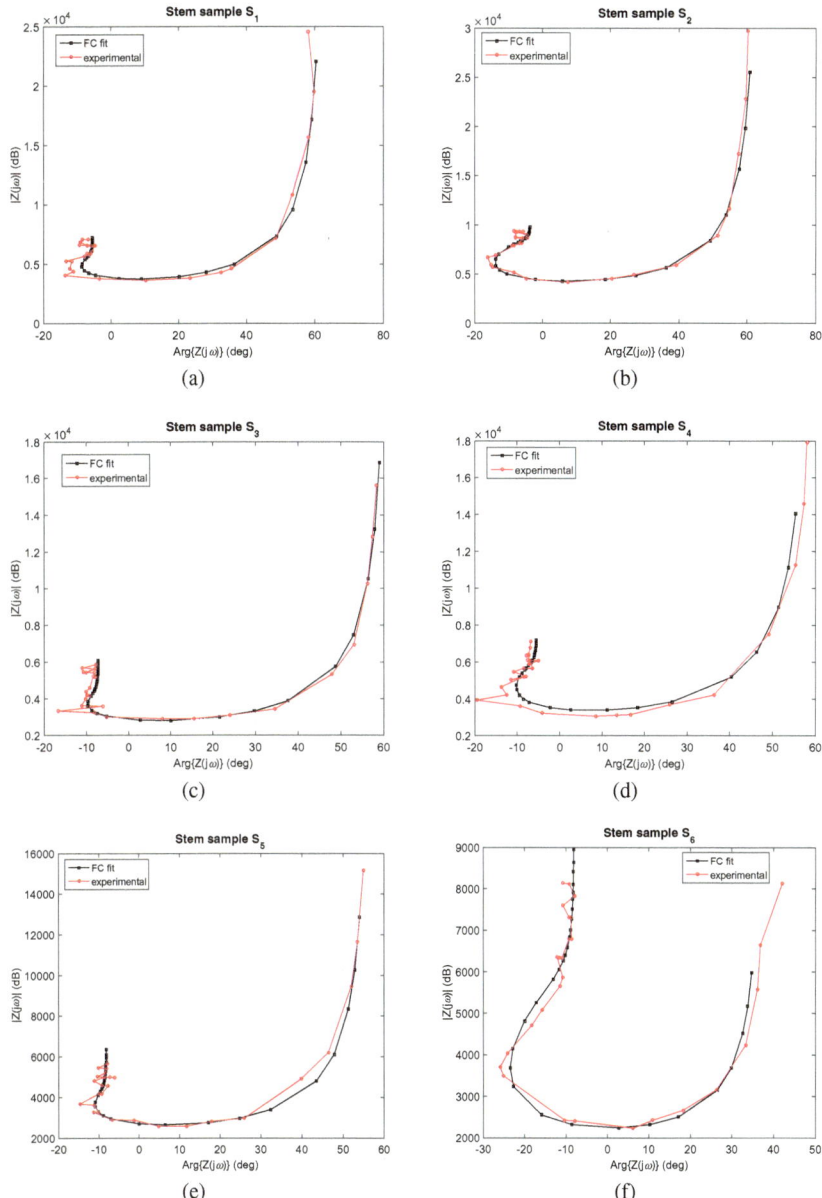

Fig. 4.3 Nichols plots for the impedance $\mathbf{Z}(j\omega)$ of the six *Hibiscus* stems in example 1

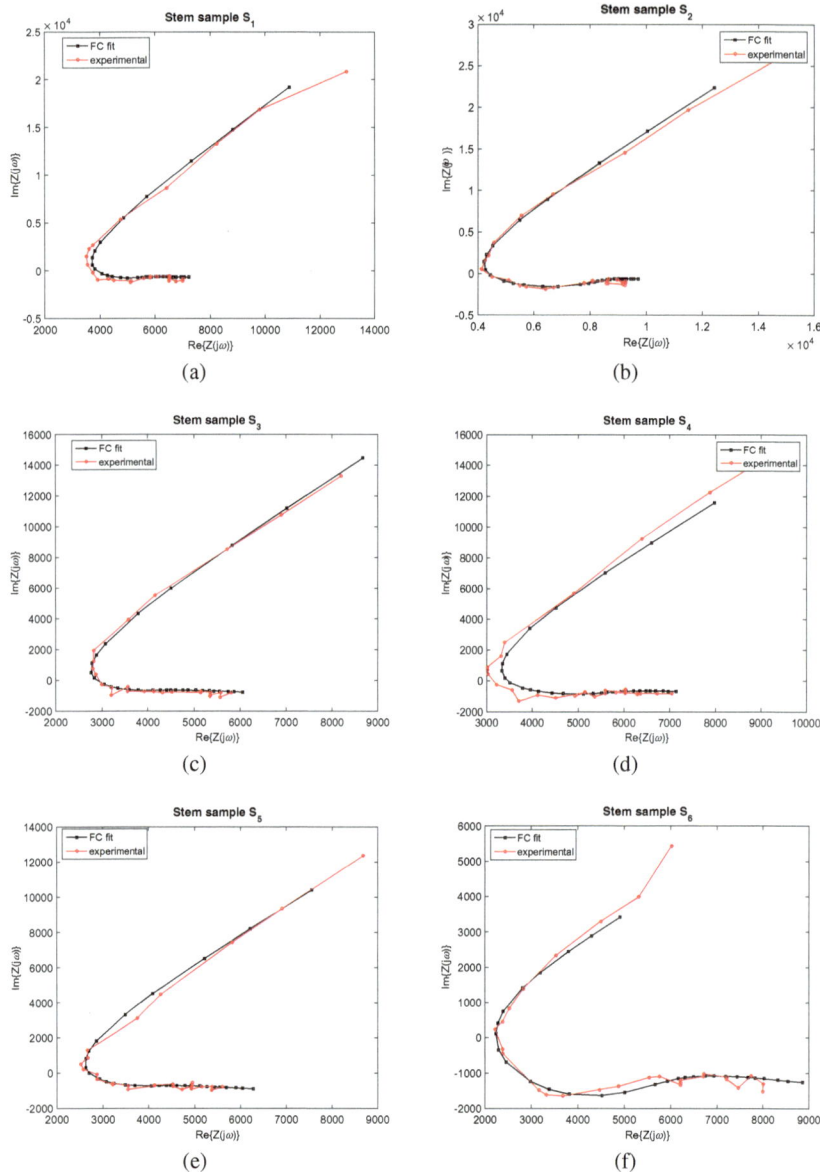

Fig. 4.4 Polar plots for the impedance $\mathbf{Z}(j\omega)$ of the six *Hibiscus* stems in example 1

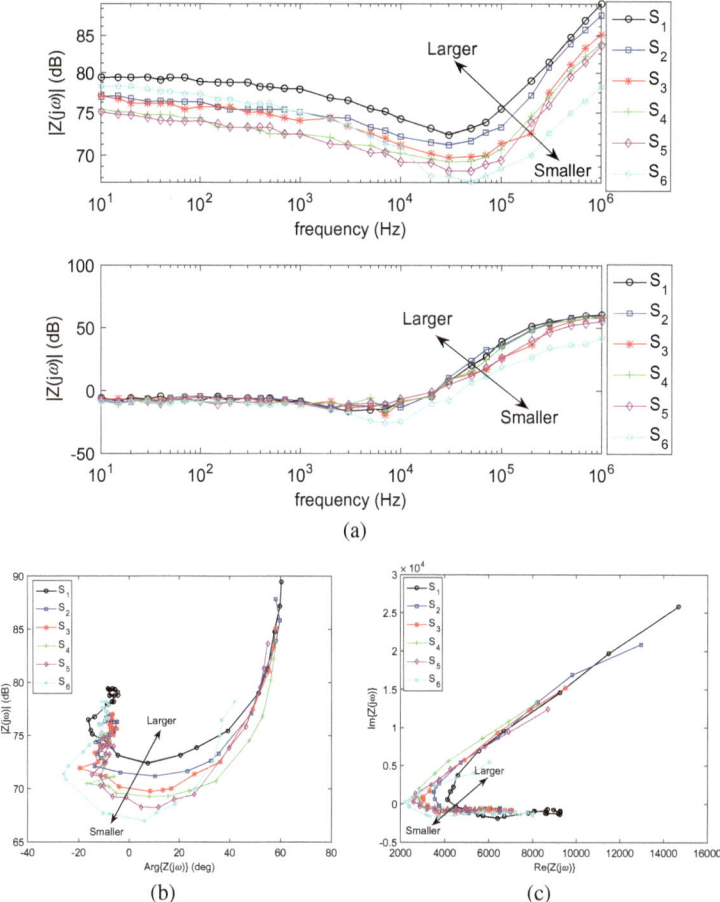

Fig. 4.5 Comparison of the impedance $\mathbf{Z}(j\omega)$ of the six *Hibiscus* stem samples in example 1: **a** Bode; **b** Nichols; **c** Polar

4.6 Clustering and Visualizing

A hierarchical clustering algorithm fed with matrix $\mathbf{M} = [c_{ij}]$, $\{i, j\} = \{1, \ldots, 12\}$ is used for visualizing possible relationships between all samples, namely the *Hibiscus* stems, fruits, and vegetables. The metric c_{ij} is based on the Canberra distance and is given by

Table 4.3 Fractional-order model parameters of the fruits and vegetables in example 2

Sample	Shape	$K(\times 10^3)$	β	$z_1(\times 10^5)$	α_1	$z_2(\times 10^3)$	α_2	J
GR		40.0	0.06	6.0	1.39	4.0	0.60	0.1461
PL		24.0	0.03	6.0	1.33	5.0	0.59	0.1568
PE		20.0	0.03	6.0	1.20	5.0	0.48	0.1856
PU		12.0	0.05	4.5	0.90	35.0	0.50	0.1542
CO		18.0	0.08	4.5	0.90	25.0	0.30	0.1159
FG		22.0	0.10	4.5	0.90	2.0	0.20	0.0686

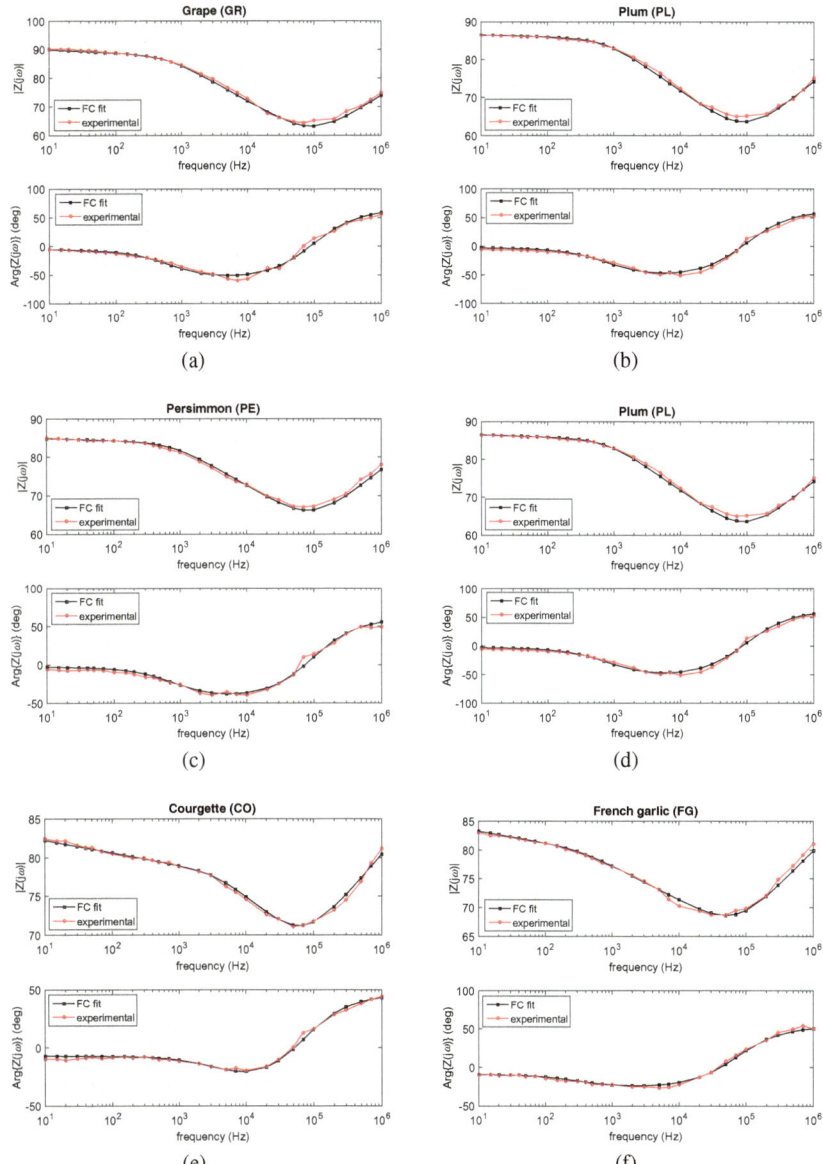

Fig. 4.6 Bode plots for the impedance $\mathbf{Z}(j\omega)$ of the six fruits and vegetables in example 2

Fig. 4.7 Visualization tree generated by a hierarchical clustering algorithm and distance c_{ij} for the *Hibiscus* stems $\{S_1, \ldots, S_6\}$, fruits $\{GR, PL, PE\}$, and vegetables $\{PU, CO, FG\}$

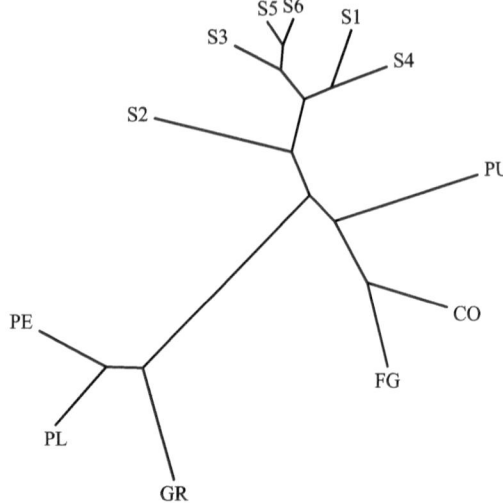

$$c_{ij} = \frac{1}{L}\sum_{k=1}^{L} \cdot \left(\frac{\left| \mathrm{Re}[\mathbf{Z}_i(j\omega_k)] - \mathrm{Re}[\mathbf{Z}_j(j\omega_k)] \right|}{\left| \mathrm{Re}[\mathbf{Z}_i(j\omega_k)] \right| + \left| \mathrm{Re}[\mathbf{Z}_j(j\omega_k)] \right|} \right) + \qquad (4.33)$$
$$\left(\frac{\left| \mathrm{Im}[\mathbf{Z}_i(j\omega_k)] - \mathrm{Im}[\mathbf{Z}_j(j\omega_k)] \right|}{\left| \mathrm{Im}[\mathbf{Z}_i(j\omega_k)] \right| + \left| \mathrm{Im}[\mathbf{Z}_j(j\omega_k)] \right|} \right),$$

where $L = 30$.

Figure 4.7 depicts the visualization tree generated by successive (agglomerative) clustering and average linkage method [30, 38]. We observe that the emerging clusters correspond to the three different groups of tissues analyzed: *Hibiscus* stems, fruits, and vegetables. These clusters are also obtained when plotting the 3D locus of the parameters $\{\alpha_1, \alpha_2, \beta\}$, (Fig. 4.8), revealing that the fractional exponents constitute a characteristic signature of the samples' dynamics.

The results demonstrate that FO models constitute simple, yet reliable models to characterize vegetable tissues. Further research directions may be the investigation of the correlations between the models parameters and anatomical and physiological characteristics of the specimens, namely freshness, water contents, physical damage, or growth stage. The potential applications of the proposed methodology will range from product storage and processing up to agriculture and technology development.

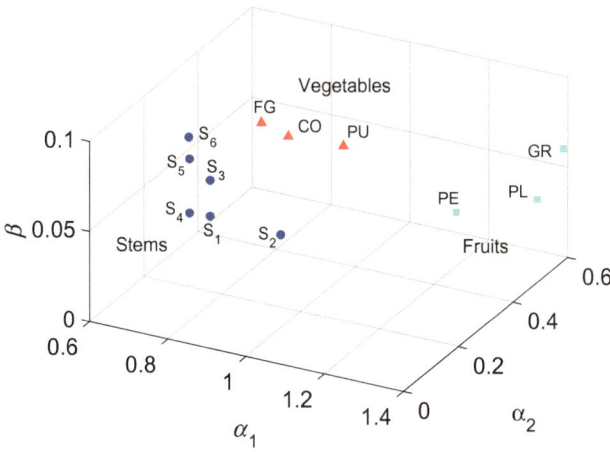

Fig. 4.8 Locus of the parameters $\{\alpha_1, \alpha_2, \beta\}$ for the *Hibiscus* stems $\{S_1, \ldots, S_6\}$, fruits $\{GR, PL, PE\}$, and vegetables $\{PU, CO, FG\}$

4.7 Conclusions

FO empirical models were adopted to describe the impedance spectra of different size stems of a plant. The experimental data was obtained by means of EIS. It was shown that a six-parameter FO transfer function represents adequately the data. The results confirmed that vegetable living matter can be thought of as natural FO devices that exhibit FO dynamics over wide frequency ranges.

References

1. Y. Ando, Y. Maeda, K. Mizutani, N. Wakatsuki, S. Hagiwara, H. Nabetani, Effect of air-dehydration pretreatment before freezing on the electrical impedance characteristics and texture of carrots. J. Food Eng. **169**, 114–121 (2016)
2. Y. Ando, K. Mizutani, N. Wakatsuki, Electrical impedance analysis of potato tissues during drying. J. Food Eng. **121**, 24–31 (2014)
3. E. Borges, A. Matos, J. Cardoso, C. Correia, T. Vasconcelos, N. Gomes, Early detection and monitoring of plant diseases by bioelectric impedance spectroscopy, in *2012 IEEE 2nd Portuguese Meeting in Bioengineering (ENBENG)* (IEEE, 2012), pp. 1–4
4. C.J.F. Böttcher, O.C. van Belle, P. Bordewijk, A. Rip, *Theory of Electric Polarization*, vol. 2 (Elsevier Science Ltd, 1978)
5. Y. Cao, T. Repo, R. Silvennoinen, T. Lehto, P. Pelkonen, Analysis of the willow root system by electrical impedance spectroscopy. J. Exp. Bot. **62**(1), 351–358 (2011)
6. A. Chowdhury, T. Bera, D. Ghoshal, B. Chakraborty, Studying the electrical impedance variations in banana ripening using electrical impedance spectroscopy (EIS), in *Third International Conference on Computer, Communication, Control and Information Technology (C3IT)* (IEEE, 2015), pp. 1–4

7. K.S. Cole, R.H. Cole, Dispersion and absorption in dielectrics I. Alternating current characteristics. J. Chem. Phys. **9**(4), 341–351 (1941)
8. D. Davidson, R. Cole, Dielectric relaxation in glycerol, propylene glycol, and *n*-propanol. J. Chem. Phys. **19**(12), 1484–1490 (1951)
9. P. Debye, Interferenz von Röntgenstrahlen und Wärmebewegung. Ann. Phys. **348**(1), 49–92 (1913)
10. P.J.W. Debye, *Polar Molecules* (Chemical Catalog Company, Incorporated, 1929)
11. P. Dejmek, O. Miyawaki, Relationship between the electrical and rheological properties of potato tuber tissue after various forms of processing. Biosci. Biotechnol. Biochem. **66**(6), 1218–1223 (2002)
12. D. El Khaled, N. Castellano, J. Gazquez, R.G. Salvador, F. Manzano-Agugliaro, Cleaner quality control system using bioimpedance methods: a review for fruits and vegetables. J. Clean. Prod. (2015)
13. T. Ellis, W. Murray, L. Kavalieris, Electrical capacitance of bean (Vicia faba) root systems was related to tissue density—a test for the Dalton model. Plant Soil **366**(1–2), 575–584 (2013)
14. S. Emmert, M. Wolf, R. Gulich, S. Krohns, S. Kastner, P. Lunkenheimer, A. Loidl, Electrode polarization effects in broadband dielectric spectroscopy. Eur. Phys. J. B **83**(2), 157–165 (2011)
15. Y. Feldman, A. Puzenko, Y. Ryabov, Non-Debye dielectric relaxation in complex materials. Chem. Phys. **284**(1), 139–168 (2002)
16. T.J. Freeborn, A survey of fractional-order circuit models for biology and biomedicine. IEEE J. Emerg. Sel. Top. Circuits Syst. **3**(3), 416–424 (2013)
17. T.J. Freeborn, B. Maundy, A.S. Elwakil, Cole impedance extractions from the step-response of a current excited fruit sample. Comput. Electron. Agric. **98**, 100–108 (2013)
18. H. Fröhlich, *Theory of Dielectrics: Dielectric Constant and Dielectric Loss* (Clarendon Press, 1958)
19. R. Garra, A. Giusti, F. Mainardi, G. Pagnini, Fractional relaxation with time-varying coefficient. Fract. Calc. Appl.Anal. **17**(2), 424–439 (2014)
20. R. Garrappa, F. Mainardi, G. Maione, Models of dielectric relaxation based on completely monotone functions. Fract. Calc. Appl. Anal. **19**(5), 1105–1160 (2016)
21. C. Greenham, Bruise and pressure injury in apple fruits. J. Exp. Bot. **17**(2), 404–409 (1966)
22. C. Greenham, K. Helms, W. Müller, Influence of virus inflections on impedance parameters. J. Exp. Bot. **29**(4), 867–877 (1978)
23. S. Havriliak, S. Negami, A complex plane analysis of α-dispersions in some polymer systems. J. Polym. Sci. Part C: Polym. Symp. **14**, 99–117 (1966) (Wiley Online Library)
24. R. Hilfer, Analytical representations for relaxation functions of glasses. J. Non-Crystal. Solids **305**(1), 122–126 (2002)
25. I.S. Jesus, J.T. Tenreiro Machado, J.B. Cunha, Fractional electrical impedances in botanical elements. J. Vib. Control **14**(9–10), 1389–1402 (2008)
26. X. Kou, L. Chai, L. Jiang, S. Zhao, S. Yan, Modeling of the permittivity of holly leaves in frozen environments. IEEE Trans. Geosci. Remote Sens. **53**(11), 6048–6057 (2015)
27. W. Kuang, S. Nelson, Dielectric relaxation characteristics of fresh fruits and vegetables from 3 to 20 GHz. J. Microwave Power Electromagn. Energy **32**(2), 115–123 (1997)
28. A. Laogun, N. Ajayi, Radio-frequency dielectric properties of some tropical african leaf vegetables. Technical report, International Centre for Theoretical Physics, Trieste (Italy) (1985)
29. S. Laufer, A. Ivorra, V.E. Reuter, B. Rubinsky, S.B. Solomon, Electrical impedance characterization of normal and cancerous human hepatic tissue. Physiol. Meas. **31**(7), 995 (2010)
30. A.M. Lopes, J.A. Tenreiro Machado, Dynamic analysis of earthquake phenomena by means of pseudo phase plane. Nonlinear Dyn. **74**(4), 1191–1202 (2013)
31. A.M. Lopes, J.A. Tenreiro Machado, Fractional order models of leaves. J. Vib. Control **20**(7), 998–1008 (2014)
32. A.M. Lopes, J.A. Tenreiro Machado, Modeling vegetable fractals by means of fractional-order equations. J. Vib. Control **22**(8), 2100–2108 (2016)
33. A.M. Lopes, J.A. Tenreiro Machado, E. Ramalho, On the fractional-order modeling of wine. Eur. Food Res. Technol. 1–9 (2016)

34. S. Mancuso, Seasonal dynamics of electrical impedance parameters in shoots and leaves related to rooting ability of olive (Olea europea) cuttings. Tree Physiol. **19**(2), 95–101 (1999)
35. B. Maundy, A. Elwakil, Extracting single dispersion Cole-Cole impedance model parameters using an integrator setup. Analog Integr. Circuits Signal Process. **71**(1), 107–110 (2012)
36. B. Maundy, A. Elwakil, A. Allagui, Extracting the parameters of the single-dispersion Cole bioimpedance model using a magnitude-only method. Comput. Electron. Agric. **119**, 153–157 (2015)
37. Y. Mizukami, K. Yamada, Y. Sawai, Y. Yamaguchi, Measurement of fresh tea leaf growth using electrical impedance spectroscopy. Agric. J. **2**(1), 134–139 (2007)
38. F. Murtagh, A survey of recent advances in hierarchical clustering algorithms. Comput. J. **26**(4), 354–359 (1983)
39. S.O. Nelson, S. Trabelsi, Factors influencing the dielectric properties of agricultural and food products. J. Microwave Power Electromagn. Energy **46**(2), 93–107 (2012)
40. R. Nigmatullin, S. Nelson, Recognition of the "fractional" kinetics in complex systems: dielectric properties of fresh fruits and vegetables from 0.01 to 1.8 GHz. Signal Process. **86**(10), 2744–2759 (2006)
41. R. Nigmatullin, Y.E. Ryabov, Cole-Davidson dielectric relaxation as a self-similar relaxation process. Phys. Solid State **39**(1), 87–90 (1997)
42. V. Novikov, K. Wojciechowski, O. Komkova, T. Thiel, Anomalous relaxation in dielectrics. Equations with fractional derivatives. Mater. Sci. Wroclaw **23**(4), 977 (2005)
43. S. Ohnishi, O. Miyawaki, Osmotic dehydrofreezing for protection of rheological properties of agricultural products from freezing-injury. Food Sci. Technol. Res. **11**(1), 52–58 (2005)
44. E.C. de Oliveira, F. Mainardi, J. Vaz Jr., Models based on Mittag-Leffler functions for anomalous relaxation in dielectrics. Eur. Phys. J. Spec. Top. **193**(1), 161–171 (2011)
45. E.C. de Oliveira, F. Mainardi, J. Vaz Jr., Fractional models of anomalous relaxation based on the Kilbas and Saigo function. Meccanica **49**(9), 2049–2060 (2014)
46. H. Ozier-Lafontaine, T. Bajazet, Analysis of root growth by impedance spectroscopy (EIS). Plant Soil **277**(1–2), 299–313 (2005)
47. U. Pliquett, Bioimpedance: a review for food processing. Food Eng. Rev. **2**(2), 74–94 (2010)
48. T. Repo, S. Pulli, Application of impedance spectroscopy for selecting frost hardy varieties of english ryegrass. Ann. Bot. **78**(5), 605–609 (1996)
49. E.C. Rosa, E.C. de Oliveira, Relaxation equations: fractional models (2015), arXiv:1510.01681
50. Sibatov, R.T., Uchaikin, D.V.: Fractional relaxation and wave equations for dielectrics characterized by the Havriliak-Negami response function (2010), arXiv:1008.3972
51. A. Stanislavsky, K. Weron, J. Trzmiel, Subordination model of anomalous diffusion leading to the two-power-law relaxation responses. EPL (Europhys. Lett.) **91**(4), 40,003 (2010)
52. M. Tiitta, L. Tomppo, H. Järnström, M. Löija, T. Laakso, A. Harju, M. Venäläinen, H. Iitti, L. Paajanen, P. Saranpää et al., Spectral and chemical analyses of mould development on Scots pine heartwood. Eur. J. Wood Wood Prod. **67**(2), 151–158 (2009)
53. K. Toyoda, R.N. Tsenkova, M. Nakamura, Characterization of osmotic dehydration and swelling of apple tissues by bioelectrical impedance spectroscopy. Drying Technol. **19**(8), 1683–1695 (2001)
54. J. Urban, R. Bequet, R. Mainiero, Assessing the applicability of the earth impedance method for in situ studies of tree root systems. J. Exp. Bot. **62**(6), 1857–1869 (2011)
55. A. Väinölä, T. Repo, Impedance spectroscopy in frost hardiness evaluation of rhododendron leaves. Ann. Bot. **86**(4), 799–805 (2000)
56. Z. Vosika, M. Lazarević, J. Simic-Krstić, D. Koruga, Modeling of bioimpedance for human skin based on fractional distributed-order modified Cole model. FME Trans. **42**(1), 74–81 (2014)
57. T. Watanabe, T. Orikasa, H. Shono, S. Koide, Y. Ando, T. Shiina, A. Tagawa, The influence of inhibit avoid water defect responses by heat pretreatment on hot air drying rate of spinach. J. Food Eng. **168**, 113–118 (2016)
58. L. Wu, Y. Ogawa, A. Tagawa, Electrical impedance spectroscopy analysis of eggplant pulp and effects of drying and freezing-thawing treatments on its impedance characteristics. J. Food Eng. **87**(2), 274–280 (2008)

59. L. Wu, T. Orikasa, K. Tokuyasu, T. Shiina, A. Tagawa, Applicability of vacuum-dehydrofreezing technique for the long-term preservation of fresh-cut eggplant: effects of process conditions on the quality attributes of the samples. J. Food Eng. **91**(4), 560–565 (2009)
60. L. XiaoHong, H. TingLin, W. GuoDong, Z. Gang et al., Effect of salt stress on electrical impedance spectroscopy parameters of wheat (Triticum aestivum l.) leaves. J. Zhejiang Univ. (Agric. Life Sci.) **35**(5), 564–568 (2009)
61. M. Zhang, T. Repo, J. Willison, S. Sutinen, Electrical impedance analysis in plant tissues: on the biological meaning of Cole-Cole α in Scots pine needles. Eur. Biophys. J. **24**(2), 99–106 (1995)
62. M. Zhang, J. Willison, Electrical impedance analysis in plant tissues: the effect of freeze-thaw injury on the electrical properties of potato tuber and carrot root tissues. Can. J. Plant Sci. **72**(2), 545–553 (1992)
63. M. Zhang, J. Willison, Electrical impedance analysis in plant tissues. J. Exp. Bot. **44**(8), 1369–1375 (1993)

Chapter 5
Future Directions

5.1 Introduction

In this chapter, we discuss possible future directions in research. Further efforts are needed to establish the specific fabrication parameters that lead to desired phase shift and extend the working lifetime of the devices. Going forward, this will lead to a substantial generalization of Ohm's law to incorporate a generalized form of nonlinear elements that have memory of their impedance state. We know that the fractional-order (FO) of some dielectrics changes with temperature, pressure, and other environmental variables. This could guide further generalizations of the fractional calculus (FC) to allow for modeling dynamic fractional-order. To date, no physically plausible mathematical model for time-varying order exists. The models that have been proposed violate causality and/or conservation of energy, among other problems. We generalize the formulation of memory functions in materials to achieve a broader generalization of the calculus of arbitrary order.

5.2 Challenges and Opportunities

The lesson learned is that power law (PL) is "the law." More and more, there is a realization that fractal geometries and FO dynamics are the norm and should no longer be considered anomalous [20]. In this chapter, we continue with the theme—fractional-order element (FOE) properties are ubiquitous and we can learn much from the diversity and robustness of nature itself. We are learning how to create new generations of fractional-order materials and devices, but it will take additional effort to open our minds to all the implications of non-integer order thinking.

There is a cultural problem that has to be addressed. During demonstrations of the fractional-order controller (FOC) discussed in Chap. 3 and Ref. [2], the common response was "We do not do it that way", and "We are used to complex digital system design and the systems seem to work, so why change to something simpler?" The

© The Author(s) 2017
R. Caponetto et al., *Fractional-Order Devices*,
SpringerBriefs in Nonlinear Circuits, DOI 10.1007/978-3-319-54460-1_5

answer to these questions should be obvious. If we kept doing things the way we "always" have, no progress would be made and there would not be any computers, let alone digital controllers. The better answer is that these designs offer a robustness to uncertainties in our system models that is unmatched in conventional techniques.

Pursuing the development of fractional-order devices (FOD) not only offers the possibility of highly useful electronic circuit elements but also allows for the study of complexity in a much broader context, see, for example, Ref. [17]. How does history (memory) of the stresses and strains on an object affect its electrical, mechanical, and other properties? As seen in Chap. 4, the study can certainly be extended to biological systems. There is no reason not to continue into financial, geologic, and other processes. In this context, developing a variety of FOD has substantial scientific as well as engineering merit since the mathematics is common to all these topics.

5.3   Achieving Specific Fractional-Order and Longer Working Lifetimes

There is still much to be done before fractance-based devices become commercially available. For example, what parameters in the various recipes lead to specific fractional-order? Can we get predictive? Are there recipes compatible with existing electronic device manufacturing processes? This compatibility may not be feasible due to conventional processes working so hard to reduce randomness and the goal in fabrication of fractances seems to depend on increased randomness.

Further work on improving the stable lifetimes of the existing fractance device recipes is critical.

Uchaikin and Sibatov provide an excellent review of fractional dynamics in various materials [19]. What more can be learned from such characterizations?

5.4   Advanced Research Opportunities

The fractional calculus allows differentiation and integration to *any* order, including complex. Does nature exhibit complex order fractional dynamic properties? There is some evidence that this may be the case. Nigmatullin has suggested a supporting theory and reported a possible case of just such a dynamical response [7–9].

5.5   Dynamic Fractance and Memfractance

We know that the impedance magnitude in a FOD can change over time and with changes in environment. We also have evidence that the order of the fractance can change with changes in the environmental conditions. There is currently no mathematical representation to incorporate time-varying FO that conserves energy and causality, that the currently popular forms for time-varying order cannot represent physical phenomena can best be demonstrated by inserting time dependence into (1.39b) and asking whether the resulting equation makes sense:

$$v(t) \overset{?}{=} \sum_{k=0}^{N-1} K(t) \frac{\Gamma(k+\alpha)}{\Gamma(\alpha)\Gamma(k+1)} i(t-k\Delta t) \left[\frac{\Delta t}{\tau(t)} \right]^{\alpha(t)} \tag{5.1}$$

with the clear implication that the amplitude scaling $K(t)$ and time scaling τ^α must change as well as the order itself. How do we interpret this? Can the K and τ values "now" apply to deep past history? Equation (5.1) makes no physical sense. In fact, this implies that the operator is no longer linear and the Laplace transform methods no longer apply. This physical reality is not covered in the literature to date.

A clue as to how to handle varying FO may be found in the evolution of memristor theory originally proposed by Leon Chua [3, 4]. His initial theoretical argument was that the operators were explicitly of integer order:

$$\delta v^{(\alpha)}(t) = m_Q \delta i^{(\beta)}(t) \tag{5.2}$$

such that

$$Z_Q(s) = m_Q s^k, \quad \text{where } k = \beta - \alpha, \tag{5.3}$$

using the notation $v^{(\alpha)} \equiv d^\alpha v(t)/dt^\alpha$, and δ suggest small-signal variations around an operating point. The parameter m_Q is set based on the large-signal history of excitation. Note that the impedance is only "locally" linear around the current operating point.

This was updated in 2014 with an article proposing the term "memfractance." It suggested "a mathematical paradigm for circuit elements with memory" [1]. Machado discussed generalizations of memristors and fractances in 2013, where he suggested using generalized impedance convertors to create a variety of possible orders given a few FOD [12]. The authors of these articles allowed the orders of the operators α and β to be non-integers, but stopped short of including dynamic variation in these orders.

The huge problem now is that you have a varying memory of the state of the device affecting the memory of the input signal. The wording here is intentional and is intended to evoke echoes of past stimulus and response that depend on prior states of the system. Is nature really this recursive in its dynamics? There is much evidence to say yes. How do we model material dynamics that exhibit $\alpha(t)$ or $\alpha(\omega)$?

In particular, can we build a test set that allows changing the FO during runs at constant frequency? How does the phase shift? Temperature variation tests with Nasicon appear to suggest that this is possible [10]. The same questions could be applied to sweeping frequency. If the dynamics are FO, then the response cannot be purely due to the input at time t, but must depend on the prior state history. Just designing experiments to illuminate these questions will be challenging, but the insights gained should apply to PL dynamics in virtually any complex system.

Given we can derive a model of dynamic FO, can we go back and look at how we might turn that into another new electronic element? Not just the static table of circuit elements envisioned by Chua, but devices that shift around that chart under dynamic control?

What if we could determine the conditions under which the FO dynamics of living tissue, from leaves to human nerve and muscle, are affected by environmental conditions? Could we then model the effects of changing drug constituents and dosages? It is now pretty clear that understanding the environmental and other conditions that result in a specific fractional-order could have significant applications in both engineering circuitry and in sensing and diagnostics. Putting these ideas together, what if we could make a system that senses environmental changes and adapts the order of the controller to compensate? This type of auto-tuning is being incorporated into integer order control systems now, but with ever-increasing complexity.

In summary, a broad research plan outline would cover a range of questions across multiple disciplines. Commonalities among various materials, including animal, vegetable, and mineral systems, could lead to a better understanding of how our world evolved the way it did.

- What are the critical factors that determine the order and bandwidth of the PL response?

  - Chemistry?
  - Fractal electrode–electrolyte interface?
  - Fractal structure of the electrolyte?

- What environmental effects change the fractional-order?
- How does the system respond during transients in the environment?
- How might we test and extend theoretical treatment?
- What are the scientific and engineering applications?

5.6 Generalizing Ohm's Law

The theory of memristance has evolved considerably. Of particular interest is Chua's statement that the fundamental circuit elements fall on the corners of the box outlined in the center of a periodic table of possible circuit elements, as shown in Fig. 5.1. The parameters α and β are used to describe the degree of memory in the "mem-conductance" device. The value of m is assumed to vary nonlinearly with the large signals applied to the device. The small-signal linear response would then appear to have order $k = \beta - \alpha$.

In memristance theory, the concept of flux as the time integral of voltage is now taken as axiomatic whether or not an actual magnetic field plays a role in the process.

Fig. 5.1 Chua's concept for a periodic table of the possible circuit elements, with symbols: R = resistor, C = capacitor, L = inductor, M = memristor, and introducing F as fractance. Note that fractance begins to fill in the line between resistance and capacitance. Do all points on the (α, β) plane represent possible electronic elements?

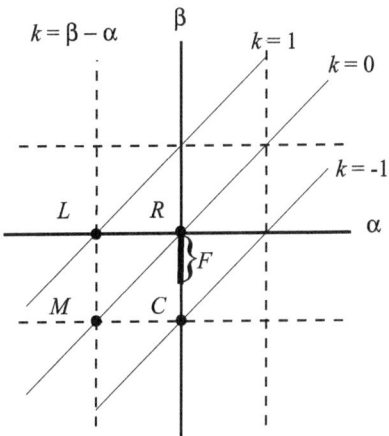

Starting with traditional components and introducing the fractor and memristor:

$$v^{(0)}(t) = m_R \, i^{(0)}(t) \qquad \text{Resistance; } m_R = R \qquad\qquad (5.4a)$$

$$v^{(0)}(t) = m_C \, i^{(-1)}(t) \qquad \text{Capacitance; } m_C = 1/C \qquad\qquad (5.4b)$$

$$v^{(-1)}(t) = m_L \, i^{(0)}(t) \qquad \text{Inductance; } m_L = L \qquad\qquad (5.4c)$$

$$v^{(0)}(t) = m_F \, i^{(\beta)}(t) \qquad \text{Fractance; } m_F = K/\tau^\beta, \, -1 < \beta < 0 \qquad\qquad (5.4d)$$

$$v^{(-1)}(t) = m_M \, i^{(-1)}(t) \qquad \text{Memristance; } m_M = M(q), \, q = \text{charge} \qquad (5.4e)$$

We can now extend the range of the exponents α and β to any real value, and potentially, any complex value, to describe an even more generalize "memfractance," allowing the evolution of the fractional-orders as well. The major difference is that now Ohm's Law is extended to highly nonlinear behavior. Nonlinearity is key to Chua's theory of how a memristance changes and then retains its state under the new conditions. As of this writing, the emphasis has been on using memristors as digital switches, while noting that the switching is actually quite slow for digital memory and computation applications. We need to get a handle on the transient behavior if we are to use such devices in analog, continuous signal operations. The difficulty in merging the theory and methods of the earlier chapters in this brief is that memfractance is assumed to be nonlinear and the methods of impedance spectroscopy are not appropriate for such systems. A more accurate question would be whether the assumption of nonlinearity is necessary? Can the methods of fractional calculus be extended to incorporate the generalization of Ohm's law to incorporate the variation in the order of the processes [11]?

The key point here is that the dynamic order is an internal parameter subject to variation in time just as any other property such as temperature. The order of a fractance is really a description of the memory properties embedded in the constitutive relation between voltage and current. We do not have direct access to the fractance

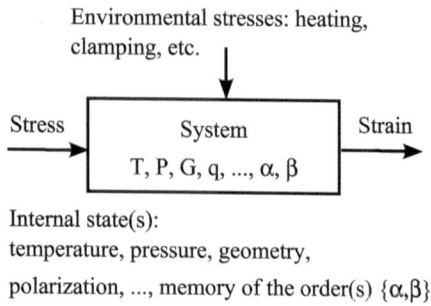

Fig. 5.2 All stress–strain relationships are affected by all of the other environmental conditions. It is our interpretation of which aspects represent the "signal of interest" versus the external environmental aspects that we try to keep constant while making measurements

order, but we can assume it varies with externally applied stresses, such as changes in concentration of constituent materials, exposure to a magnetic field, or changes in temperature or pressure. This idea is shown schematically in Fig. 5.2, highlighting the difference between external and internal variables. Expanding the constitutive relation,

$$v^{(\alpha(t))} = m_Q(\mathbf{x}, i, t)i^{(\beta(t))}(t) \tag{5.5}$$

with the internal states $\mathbf{x}$ evolving according to

$$\dot{\mathbf{x}} = f(\mathbf{x}, \mathbf{w}, i, t), \tag{5.6}$$

where $m_Q(\mathbf{x}, i, t)$ represents the dynamically evolving impedance, which now actually represents a memory function operator. The externally applied environment variables that we have access to are represented by $\mathbf{w}$ and the internal system variables are represented by $\mathbf{x}$ which includes the order parameters α and β. The history of the evolution of the orders α and β may complete the description of "memory of memory" outlined by Lorenzo and Hartley in [6].

The meaning of, and how to compute, the evolution of the constitutive equation (5.5) is not yet clear. As pointed out above, the GL and RL forms no longer apply to time-varying order conditions. How does the memory of the environment contribute to the weighting of memory of the signal? There is no reason not to turn the question around and treat any other environmental variable as the "signal" and treat the applied current as part of the "environment." Nature does not care how we name things. This symmetry may be one of the major clues as to develop a new theory. When one looks at mixed phenomena, such as piezoelectric effects, the problem has to be restated in terms of tensors. Herrmann made a first suggestion for defining fractional-order tensor operators, but how to apply this to electromechanical constitutive relationships is still to be addressed [5].

Some hints as to a new form of the nonlinear fractional-order operator may come from the realization that the exponent α in dielectric relaxation is always between zero and one. When one looks at the integral form of the Grünwald–Letnikov sum, we can imagine a revision to the recursion formula for the computation of the weights w_j. The nonlinear form would most likely not be of the form of a convolution integral, but the weights would be probably be associated with a moment in time. Each sample in past history would be weighted by a diminishing factor depending on the state of the system. An idea is that all of the previous weights might be updated each time step by multiplying a factor associated with "current" time. If a weight is decimated to near zero, such as might happen if a capacitive device were heated to the point that it become conductive, all of the old weights would be reduced to near zero, erasing the "memory" of the charge stored. Once the weight associated with some time in the past is fully decimated, it would never recover. Thus, cooling the capacitor back to being a good dielectric would not restore the past charge value.

Additional ideas for generalizing electronic circuit theory have been proposed in references such as [14, 15, 18]. In particular, extensions of the Cole–Cole model discussed in Chap. 1, other forms of curve fits to dielectric measurements have been proposed. Some of these were used in the analysis of impedance measurements for vegetable matter in Chap. 4. These suggest further extensions of the fractional calculus using matrix methods. Possibilities abound and it is now clear that extensive dynamical testing of electronic and mechanical properties will be required to guide further development of the mathematics for use in physical modeling.

As noted, there are many possible definitions for fractional-order operators. We are guided here by the quest for the form(s) that best describe physical processes. Only by experiment can we make progress on this quest.

5.7 Teaching Fractional Calculus and Its Applications

Physics and mathematics texts have been including nonlinear dynamics and chaos for some time. There is not that much of a leap to including non-local interactions into the description of dynamics. As people get ever more comfortable with long-range human interactions over the Internet, they may be more adaptable to the idea of non-local interactions at the micro- and mesoscopic scales.

Experience with teaching advanced undergraduate students by Bohannan has shown that the earlier the ideas of fractional calculus are introduced, the easier it is to internalize the concepts. Students are astonished that one equation, the Grünwald–Letnikov limit sum form, describes both differential and integral operations, was invented during the period of the U.S. civil war. The overwhelming response has been "Why was this not introduced in my calculus course?"

Resistance to incorporating FC into the curriculum can be summarized and addressed along the following lines:

- Fractional calculus is too hard and not appropriate for undergraduates. This might be debatable, but pretty much any new concept can be difficult at first. Introducing real physical evidence for PL dynamics makes the ideas much more intuitive.

- Fractional calculus problems are not included in any qualifying examinations, such as the Graduate Record Exam (GRE) in the U.S. This might be the most difficult paradigm to overcome, but might be the most important to challenge. Should we "teach to the test"?
- Problems can be solved without the complexity of the fractional calculus. This argument is often coupled with the comment that digital controllers can be arbitrarily complex, so why introduce the complicated thinking associated with fractional-order operators? The answer to this is that once you understand fractional calculus, complex dynamics arising in many distinct and seemingly disparate arenas become quite a bit more connected.

Approaches to introducing fractional calculus vary, but introducing the integral forms of Riemann and Liouville may not be appropriate for the initial exposure. The Grünwald–Letnikov form using discrete measurements converges quite rapidly for most physical simulations. Introducing the recursive weighting of (1.19) using a spreadsheet application or a programming environment such as Python that allows plotting can be very effective at developing an intuition for the history dependence implied by a fractional-order operator. Coupling that with rigorous attention to dimensional consistency exposes the need for the time-scaling parameter τ in (1.13). Only by introducing the fractional calculus to a broad range of naive minds can we hope to get the next step forward in understanding the true beauty of the complexity of the universe. The authors do not consider this an overstatement.

Most of the material published on the connections among impedance spectroscopy, circuit design, and fractional-order control is written at a very scholarly level. A great deal more effort needs to be made to generate introductory material for advanced undergraduates. This might be initiated by the development of a student supplement to this monograph available on the Internet. Included might be a repository for fractance element recipes for chemistry and materials science students, design guides for developing fractional-order circuits, example problems in fractional equations, simulations of fractional dynamics, curve-fitting schemes for analysis of impedance data, and template code sections in MATLAB, Python, or MAPLE, just to suggest a few. Some examples of this kind of material are already available on the Web, but scattered and difficult to discover.

5.8 Conclusion

Linking physical experiments with the mathematics may be the best way to overcome the difficulties with the fractional-order unit problems now plaguing the basic definitions of fractional calculus. Development of additional fractional-order devices and their physical demonstration will accelerate the acceptance of fractional calculus and with that extend its use. In just this one case of the fractional-order device we demonstrate that the ideal low-loss capacitor may its place in switching and other applications, but it is not the ideal component for all occasions.

We have attempted to pull together evidence from a broad range of scientific and engineering specialties that the study of fractional-order devices is justified on many accounts.

- Such devices have numerous potential applications.
- Developing such devices leads to better understanding of complex systems.
- Such investigations push the boundaries of our understanding of the generalized mathematics of fractional calculus and point to how to interpret the results.
- This could lead to an ultimate generalization of Ohm's Law.

Once considered paradoxical, we now see that the fractional calculus allows the construction of highly accurate models of physical dynamics. The implication that dynamics are driven by long-range interactions in time and space is reflected directly in the structure of the mathematical formalism. We are now turning that around to think of materials exhibiting fractional dynamics into devices that can act as mathematical operators. We have only begun to explore the implications and possibilities for applications. See, for example, the notes from brainstorming sessions on the direction of research into the fractional calculus and its applications [13, 16].

It took almost a century for Newton's principles of calculus and physics to be accepted into general usage. After more than three centuries from the first musings of L'Hopital and Leibnitz, the ideas of the generalized calculus of arbitrary order are now being accepted and applied to an ever expanding range of applications, factional-order devices being one of the most promising.

References

1. M.-S. Abdelouahab, R. Lozi, L.O. Chua, Memfractance: a mathematical paradigm for circuit elements with memory. Int. J. Bifurcat. Chaos **24**(9), 1430023(29) (2014)
2. G.W. Bohannan, Analog fractional-order controller in temperature and motor control applications. J. Vib. Control **14**(9–10), 1487–1498 (2008)
3. L.O. Chua, S.M. Kang, Memristive Devices Syst. **64**(2), 209–223 (1976)
4. L.O. Chua, Nonlinear circuit foundations for nanodevices, part I: the four-element torus. Proc. IEEE **91**(11), 1830–1859 (2003)
5. R. Herrmann, *Fractional Calculus: An Introduction for Physicists*, 2nd edn. (World Scientific, NJ, 2014)
6. C.F. Lorenzo, T.T. Hartley, Variable order and distributed order fractional operators. Nonlinear Dyn. **29**, 57–98 (2002)
7. R.R. Nigmatullin, A.L. Mehaute, Section 10. Dielectric methods, theory and simulation: is there geometrical/physical meaning of the fractional integral with complex exponent? J. Non–Cryst. Solids **351**, 2888–2899 (2005)
8. R.R. Nigmatullin, Theory of dielectric relaxation in non-crystalline solids: from a set of micro-motions to the averaged collective motion in the mesoscale region. Phys. B. **358**, 201–215 (2005)
9. R.R. Nigmatullin, Fractional kinetic equations and universal decoupling of a memory function in mesoscale region. Phys. A. **363**, 282–298 (2006)
10. P. Pasierb, S. Komornicki, R. Gajerski, S. Koziński, P. Tomczyk, M. Rekas, Electrochemical gas sensor materials studied by impedance spectroscopy, Part I: Nasicon as a solid electrolyte. J. Electroceram. **8**(1), 49–55 (2002)

11. H. Sun, H. Sheng, Y.-Q. Chen, W. Chen, On dynamic–order fractional dynamic system, arXiv:1103.0082v2 [math–ph], (10 pp.) (2011)
12. J.A. Tenreiro Machado, Fractional generalization of memristor and higher order elements. Commun. Nonlinear Sci. Numer. Simulat. **18**, 122–246 (2013)
13. J.A. Tenreiro Machado, F. Mainardi, V. Kiryakova, Fractional calculus: quo vadimus? (where are we going?) Contributions to round table discussion held at ICFDA 2014. Fractional Calc. Appl. Anal. **18**(2), 495–526 (2015). doi:10.1515/fca-2015-0031
14. J.A. Tenreiro Machado, Fractional-order junctions. Commun. Nonlinear Sci. Numer. Simul. **20**(1), 1–8 (2015), http://dx.doi.org/10.1016/j.cnsns.2014.05.006
15. J.A. Tenreiro Machado, Matrix fractional systems. Commun. Nonlinear Sci. Numer. Simul. **25**(1–3) 10–18 (2015). doi:10.1016/j.cnsns.2015.01.00
16. J.A. Tenreiro Machado, F. Mainardi, V. Kiryakova, T. Atanackovic, Fractional calculus: do venons-nous? Que sommes-nous? O allons-nous? (Contributions to round table discussion held at ICFDA 2016). FCAA—Fractional Calc. Appl. Anal.—Int. J. Theory Appl. **19**(5), 1074–1104 (2016). doi:10.1515/fca-2016-0059
17. J.A. Tenreiro Machado, Bond graph and memristor approach to DNA analysis. Nonlinear Dyn. (2016). doi:10.1007/s11071-016-3294-z
18. J.A. Tenreiro Machado, A.M. Galhano, Generalized two-port elements. Commun. Nonlinear Sci. Numer. Simul. **42**, 451–455 (2017)
19. V. Uchaikin, R. Sibatov, *Fractional Kinetics in Solids: Anomalous Charge Transport in Semiconductors, Dielectrics, and Nanosystems* (World Scientific, Singapore, 2013)
20. B.J. West, *Fractional Calculus View of Complexity: Tomorrow's Science* (CRC Press, Boca Raton, 2016)